UNDERSTANDING PHILOSOPHY OF SCIENCE

'This is the best introduction to philosophy of science I have read. I will certainly use it. The writing is wonderfully clear without being simplistic. It is not at all too difficult for second and third year students. Many of my philosophy of science students have no background in philosophy, and I'm sure they will find the book accessible, informative, and a pleasure to read. I read this manuscript with my students in mind. This is the book we've been looking for.'

Peter Kosso, *Northern Arizona University*

Few can imagine a world without telephones or televisions; many depend on computers and the Internet as part of daily life. Without scientific theory, these developments would not have been possible.

In this exceptionally clear and engaging introduction to the philosophy of science, James Ladyman explores the philosophical questions that arise when we reflect on the nature of the scientific method and the knowledge it produces. He discusses whether fundamental philosophical questions about knowledge and reality might be answered by science, and considers in detail the debate between realists and antirealists about the extent of scientific knowledge. Along the way, central topics in the philosophy of science, such as the demarcation of science from non-science, induction, confirmation and falsification, the relationship between theory and observation, and relativism, are all addressed. Important and complex current debates over underdetermination, inference to the best explanation and the implications of radical theory change are clarified and clearly explained for those new to the subject.

The style is refreshing and unassuming, bringing to life the essential questions in the philosophy of science. Ideal for any student of philosophy or science, this book requires no previous knowledge of either discipline. It contains the following textbook features:

- suggestions for further reading
- cross-referencing with an extensive bibliography.

James Ladyman is Lecturer in Philosophy at the University of Bristol, UK.

UNDERSTANDING PHILOSOPHY OF SCIENCE

James Ladyman

London and New York

First published 2002
by Routledge
11 New Fetter Lane, London EC4P 4EE

Simultaneously published in the USA and Canada
by Routledge
29 West 35th Street, New York, NY 10001

Reprinted 2003

Routledge is an imprint of the Taylor & Francis Group

Typeset in Sabon by RefineCatch Limited, Bungay, Suffolk
Printed and bound in Great Britain by
Biddles Ltd, Guildford and King's Lynn

British Library Cataloguing in Publication Data
A catalogue record for this book is available
from the British Library

Library of Congress Cataloging in Publication Data
Ladyman, James, 1969– .
  Understanding philosophy of science/James Ladyman.
    p. cm.
  Includes bibliographical references and index.
  1. Science – Philosophy. I. Title.
Q175.L174 2001
501 – dc21   2001048105

ISBN 0–415–22156–0 (hbk)
ISBN 0–415–22157–9 (pbk)

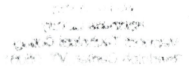

For Audrey Ladyman

Contents

—⁘—

CONTENTS

CONTENTS

Preface

———•⊙•———

This book is intended to provide an introduction to the philosophy of science. In particular, it is aimed at science students taking a philosophy of science course but no other philosophy classes, as well as at those students who are studying philosophy of science as part of a philosophy degree. Hence, I have assumed no prior knowledge of philosophy, and I have not relied upon detailed knowledge of science either. I have also avoided using any mathematics. This means that some issues are not discussed despite their interest. For example, the implications of quantum mechanics for philosophy of science, and the mathematical theory of probability and its use in modelling scientific reasoning are not dealt with here. Nonetheless, an introductory text need not be superficial and I have tried to offer an analysis of various issues, such as induction, underdetermination and scientific realism from which even graduate students and professional philosophers may benefit. My aim throughout has been to make the reader aware of questions about which they may never have thought, and then to lead them through a philosophical investigation of them in order that they appreciate the strength of arguments on all sides, rather than to offer my own views. Hence, there are few answers to be found in what follows and if my readers are left puzzled where previously they were comfortable then I will be satisfied.

I hope this book will also interest scientists and general readers who are curious about the philosophy of science. I have tried to keep the exposition clear and accessible throughout, and also to illustrate important lines of argument with everyday and scientific examples. However, the reader will find that the discussion in Chapter 5 is largely about the historical and philosophical background to the contemporary debate about scientific realism. Those who do not see its relevance immediately are urged to persevere, since the issues discussed are of fundamental importance. Finally, I must confess in advance to historians that I have subordinated historiography to my pedagogical aims by sometimes presenting a narrative that only just begins to address the complexities and ambiguities of the historical development of philosophy and science.

For the benefit of the reader, the first instances of each term expanded in the Glossary are set in bold.

Acknowledgements

———o☉o———

I am very grateful to all those who have taught me about philosophy of science, but special thanks are due to Richard Francks, Steven French, David Papineau, Anthony Sudbery and Bas van Fraassen. Thanks to all my students, especially Carlton Gibson and Nick Talbot who gave me feedback on portions of this book at early stages. Katherine Hawley kindly allowed me to see her lecture notes on Kuhn, which were very helpful. Dawn Collins, Andrew Pyle, Paul Tappenden and an anonymous referee read the whole manuscript and gave me extensive comments for which I am extremely grateful. Thanks also to Leah Henderson who gave me useful advice on the later chapters, and Jimmy Doyle who eventually managed to read the preface and the acknowledgements. I would like to thank all my colleagues in the Department of Philosophy at the University of Bristol for affording me study leave with which to write this book, and also the *Arts and Humanities Research Board* for a research leave grant. I am very grateful to Tony Bruce, philosophy editor at Routledge, who encouraged me to undertake this project, and to Siobhan Pattinson who guided it through the publication process. Finally, thanks to Audrey and Angela Ladyman for their love and support.

Chapter 8 includes material reprinted from *Studies in History and Philosophy of Modern Physics*, vol. 28(3), French, S. and Ladyman, J., 'Superconductivity and structures: revisiting the London account', pp. 363–393; copyright (1997), with permission from Elsevier Science. Also, *Studies in History and Philosophy of Science*, vol. 29(3), Ladyman, J., 'What is structural realism?', pp. 409–424; copyright (1998), with permission from Elsevier Science.

James Ladyman

Introduction

———·◯·———

In many ways, our age is no different from any other: most people work hard merely to survive, while a few live in the lap of luxury; many perish in wars and conflicts, the causes of which they have no control over; the cycle of birth, reproduction and death is fundamentally the same for us as it was for our distant ancestors. Yet certain features of the contemporary world are quite new: for example, I can pick up the phone and speak to a relative on the other side of the globe, and I can see that it is indeed a globe that I inhabit by looking at a photograph taken from space; many people's everyday lives are enhanced by, and unimaginable without, computers, televisions and music systems; medicine can treat forms of illness and injury that would have brought certain death for earlier generations. On the downside, but equally unprecedented, the nuclear weapons that many countries now have are sufficient in number to wipe out almost all life on the planet, and our skies and oceans are polluted by substances that only exist because we make them in chemical factories.

Whether good or bad in their effects, none of these technologies would exist without science. It is possible to develop ploughs, wheels, bandages and knives without much in the way of theory, but without the scientific theories and methods developed mainly in the last few hundred years there would be no electronic devices, spacecraft, micro-surgery or weapons of mass destruction. The products of science and technology have a huge effect on the way we live our lives and how we shape our environment; if you are in any doubt about this try and imagine going through an average day without using anything powered by electricity or containing plastic.

The importance of science does not only derive from its use in technology. Science enjoys unparalleled prestige in society compared with other institutions, and everyone is likely to agree about the need to fund and understand modern science while many may deride modern art or literature. Furthermore, most people are likely to trust the word of a scientist much more than they do that of a journalist, lawyer or politician (although that may not be saying much). Rightly or wrongly, science is often thought to be the ultimate form of objective and rational inquiry, and scientists are widely regarded as being able to gather and interpret evidence and use it to arrive at conclusions that are 'scientifically proven' and so not just the product of ideology or prejudice. Courts do not convict or acquit someone of a crime on the say-so of a priest or a novelist, but they do routinely rely to large extent on the evidence of an expert witness who is a scientist of some sort; if a ballistics expert says that a bullet came from a certain direction, or a pathologist says that a person had a certain drug in their system when they died, their testimony will usually be taken as establishing the facts of the case. Most of us consult a doctor when we have something wrong with us and if the doctor prescribes some drug or other therapy we take it assuming that it will help with our symptoms and not itself cause us harm. Often, modern medicine is explicitly claimed to be 'evidence-based' and hence scientific. Similarly, if the scientists appointed by the government say that a particular food or chemical is unsafe, its use and sale will be banned.

The examples above concerning justice, health and safety could readily be expanded to cover activities from engineering and construction to fishing and farming. Hence, in almost all areas of modern life, people are likely to seek or rely indirectly upon the scientific evidence and the opinions of scientists before making important decisions. Whether or not we as individuals share this faith in science and scientists, our lives are enormously affected by it, and this is one reason why understanding and thinking about science is important. Of course, most of us know very little science, and the degree of specialisation within particular sciences is now so great that no individual could possibly know all there is to know about any one scientific field let alone all about science in its entirety. For this reason, we have no choice but to rely upon co-operation and co-ordination between many individuals in order to develop further and apply scientific thought. However, there are some features of science that are more or less universal and which we can investigate philosophically without needing to know much about the cutting edge of scientific research.

Before thinking about what philosophy of science is about, it will be helpful to say what it is not about. Obviously, there are important ethical questions raised by scientific research, such as whether it is morally acceptable to conduct experiments on animals that cause them suffering, or to give psychiatric patients treatments when they may be incapable of giving their informed consent. Similarly, there are important social, political and economic questions about what research to fund and what not and, for example, whether or not to build nuclear power stations, and whether the genetic engineering of plants and animals is ethical or practically advisable. Although science policy and the ethics of scientific research ought to be informed by the philosophy of science, and indeed are part of the philosophy of science broadly conceived, they are not addressed here. Furthermore, as philosophers, we are not primarily concerned to make progress in any of the particular sciences (although philosophical thinking has often affected how work in the particular sciences is carried out and philosophical inquiry sometimes overlaps with theoretical science).

While there are other disciplines that study the sciences, the types of questions they address and their means of trying to answer them are different from those in the philosophy of science. Questions about, for example, the development of particular scientific disciplines and theories need to be addressed by historians of science, not philosophers. On the other hand, questions like, 'what sort of personality makes for a good scientist?' or 'what role do journals play in the communication and assessment of theories in physics?' are matters for the psychology or sociology of science, respectively. Philosophical questions about science, like philosophical questions in general, cannot be answered by going out in the world and gathering information, and finding out what happened, or how a particular scientific community is, as a matter of fact, organised; rather, philosophical inquiry proceeds by analysis, argument and debate.

This characterisation of history, sociology and psychology as empirically based and distinct in both subject matter and method from philosophy is itself philosophically controversial. Many philosophers think that the traditional conception of philosophy as a subject based on armchair reflection is untenable and that philosophy is really continuous with empirical inquiry and science itself (this view is known as *naturalism*). On this view, questions about scientific methodology and knowledge in philosophy of science are really continuous with questions in cognitive science about how human beings reason and form beliefs. However, one need not imagine an absolute distinction between philosophy and empirical forms of inquiry to appreciate the broad differences between the latter and the study of philosophical questions that arise when we reflect on science.

Of course, this characterisation is of little use unless we know what science is, so perhaps the most fundamental task for the philosophy of science is to answer the question, 'what is science?'. Given the status of science, this question is of great importance and many philosophers have sought to provide an answer so that it can be used to assess whether beliefs that are claimed to be scientific really are. The problem of saying what is scientific and what is not is called the **demarcation problem**. Some people have claimed scientific status for beliefs and practices, such as those of astrology, creationism (the doctrine that God created the Earth a few thousand years ago as stated in the Bible), Marxism and psychoanalysis, and some philosophers have wanted to be able to show that they are not scientific, that they are in fact merely pseudo-scientific. It is usually thought that if there is anything of which science consists it is a method or set of methods, so the study of scientific method (known as **methodology** of science) is at the centre of the philosophy of science.

We may not yet know how to define science or how to tell whether certain contentious activities or beliefs count as scientific or not, but we certainly have lots of examples of sciences. It is usual to divide the sciences into two types, namely the natural sciences and the social sciences. The former have as their object of study the natural world and include physics, chemistry, astronomy, geology and biology; the latter study the specifically human or social world and include psychology, sociology, anthropology and economics. Because the social sciences study the behaviour and institutions of human beings, they must deal with meanings, intentional actions and our apparent free will; hence, the philosophical questions they raise are often quite different from those raised by the natural sciences. Furthermore, it is an important issue in the philosophy of the social sciences whether or not a subject such as sociology is, can, or should be, scientific. Such questions do not arise for the natural sciences – if anything is a science then physics certainly is. For the purposes of this book (and here I follow standard practice) the philosophy of science is the philosophy of natural science, although many of the topics to be discussed are of concern in the philosophy of social science as well.

Philosophy of science as epistemology and metaphysics

Apart from any philosophical interest that we may have in science because of its status and influence on our lives, science is important to philosophy because it seems to offer answers to fundamental philosophical questions. One such question is 'how can we have knowledge as opposed to mere belief or opinion?', and one very general answer to it is 'follow the scientific method'. So, for example, whatever any of us may believe, rightly or wrongly, about whether smoking causes cancer or traffic fumes cause asthma, a government will not act unless there is scientific evidence supporting such beliefs (of course, they may still not act even when there is evidence). Similarly, in all the examples mentioned above, respect is accorded to the views of scientists because their conclusions are supposed to have been reached on the basis of proper methods of gathering and assessing evidence, and hence are supposed to be justified.

The branch of philosophy that inquires into knowledge and justification is called **epistemology**. The central questions of epistemology include: what is knowledge as opposed to mere belief?; can we be sure that we have any knowledge?; what things do we in fact know?. The first of these is perhaps the most fundamental epistemological question. Each of us has many beliefs, some true and some false. If I believe something that is, as a matter of fact, false (suppose, for example, that I believe that the capital city of Australia is Sydney) then I cannot be said to know it. In logical terminology we say a **necessary condition**, that is a condition that must be satisfied, for somebody knowing some **proposition** is that the proposition is true. In other words, if somebody knows some proposition then that proposition is true. (The converse obviously does not hold; there are lots of propositions that are true but which nobody knows, for example, there is a true proposition about how many leaves there are on the tree outside my window, but I presume nobody has bothered to find out what it is.) Where someone believes something that turns out to be false (no matter how plausible it seemed) then we would say that they thought they knew it but that in fact they did not.

Suppose too that another necessary condition for somebody knowing some proposition is that he or she believes that proposition. We now have two necessary conditions for knowledge; knowledge is at least true belief, but is that enough? Consider the following example: suppose that I am very prone to wishful thinking and every week I believe that my numbers will come up on the lottery, and suppose that one particular week my numbers do in fact come up; then I had a belief, that my numbers would come up, and it was a true belief, but it was not knowledge because I had no adequate reason to believe that my numbers would come up on that particular week rather than on all the other weeks when I believed they would come up, but when they did not. Hence, it may be the case that I believe something, and that it is true, but that I do not know it.

So it seems that for something someone believes to count as knowledge, as well as that belief being true, something else is required. My belief about the lottery in the example above did not count as knowledge because I lacked an adequate reason to believe that I would win that week; we would say that my belief was not justified. The traditional view in epistemology has been that knowledge can only be claimed when we have an adequate justification for our beliefs, in other words, knowledge is *justified* true belief. Although recently this 'tripartite' definition of knowledge has been the subject of much criticism and debate, justification is still often regarded as necessary for knowledge. This brings us to the issue of what justification amounts to and, as suggested above, justification is often thought to be provided by following scientific methods for testing or arriving at our beliefs (the word science comes from the Latin word *scientia*, which means knowledge).

So one area of philosophy that overlaps considerably with philosophy of science is epistemology. The epistemological questions that are addressed in later chapters (along with some of the competing answers to them) include the following. What is the scientific method? How does evidence support a theory? Is theory change in science a rational process? Can we really be said to know that scientific theories are true?

If we accept the idea that science really does give us some sort of knowledge then we must examine what scientific theories tell us about how the world is, and decide what is the scope of scientific knowledge. The modern scientific picture of the world seems to tell us a great deal, not just about how things are now, but how they were millions and even billions of years ago. Astrophysics tells us about the formation of the Earth, the solar system and even the universe, geophysics tells us about the development of mountains, continents and oceans, and biochemistry and evolutionary biology tell us about the development of life itself. Such scientific theories tell us more about familiar things, so, for example, we may learn where a particular river used to flow or how bees pollinate flowers. However, scientific theories, especially those in physics and chemistry, also describe entities that are not part of our everyday experience, such as molecules, atoms, electromagnetic waves, black holes, and so on. Such theories raise particular problems and questions in the philosophy of science; for example, should we believe in the existence of such esoteric and unobservable entities, and if so, what is to count as evidence for their existence and how do we manage to refer to them?

Of course, science does not just describe the world; it also gives us explanations of how and why things are as they are. Often this involves describing unobservable causes of things we observe. Hence, Newton is not famous for discovering that unsupported objects fall to the Earth, he is famous for explaining why they do so (the gravitational force is what causes apples to fall out of trees), and for giving us a law that allows us to calculate the rate at which they do so. Newton's mechanics, like many scientific theories, is formulated in terms of a few fundamental principles or laws. Central to our understanding of science is this idea of *laws of nature*; for example, it is supposed to be a law of nature that all metals expand when heated. So science seems to tell us about the ultimate nature of things, what the world is made of and how it works. It has even been thought that science has replaced **metaphysics** not just by telling us about what exists, and explaining what happens in terms of laws of nature and causation, but also by answering other fundamental philosophical questions about, say, the nature of space and time. But what exactly is a law of nature, and what does it mean to say that something has caused something else? What is it to explain something?

Many philosophers and scientists take it for granted that the aim of science is not merely to describe what we see, but also to arrive at the truth about the unobservable entities, laws and causes that lie behind the phenomena we observe. On the other hand, there is also a long tradition of disregarding questions about the real nature of things, the laws of nature and so on, and emphasising instead the search for theories that accurately predict what can be observed, without worrying about whether they are true or false beyond that. The question on which this book will focus is, 'ought we to believe in the unobservable entities postulated by our best scientific theories?', or more crudely, 'do electrons really exist?'. You might think this question makes little sense because electrons are, in fact, observable. After all, don't television sets work by firing electrons at a phosphorus screen, and so don't we, indirectly at least, observe electrons all the time? Exactly what is meant by observability will be discussed in the latter part of Chapter 6; however, it ought to be clear that electrons, atoms and the like are not observable in the same way that tables and trees are. Scientific realism is the view that we should believe in the likes of electrons, whereas scientific antirealism is the view that we should stop short of believing in the truth of scientific theories and content ourselves with believing what they say about what we can observe. In trying to decide the issue of scientific realism we will have to address all the epistemological and metaphysical questions mentioned above along the way.

Part I

The scientific method

1

—∘⊖∘—

Induction and inductivism

1.1 The sceptic's challenge

Our starting point is the desire to arbitrate the following dispute that arises when Alice, who has been reading *A Brief History of Time* by Stephen Hawking, is trying to explain the exciting things she has learned about the Big Bang and the history of the universe to her friend Thomas.

—∘⊖∘—

Alice: ... and so one second after the Big Bang the temperature of the universe was about ten thousand million degrees, which is about the same as the temperature in the middle of the explosion of a nuclear bomb.

Thomas: Do you really buy all that stuff? Don't you think it's a bit far-fetched?

Alice: Of course I believe it, and I don't think it is any more far-fetched than the fact that this table we are sitting at is almost all empty space and that it is made of atoms so tiny that millions of them could fit on the end of a pin.

Thomas: Exactly, it is just as far-fetched and you are just gullible for believing it.

Alice: But that is what science tells us.

Thomas: 'Science' doesn't tell us anything; scientists, people like you or me, tell us things and like all people they tell us what is in their interest to tell us.

Alice: What do you mean?

Thomas: Isn't it obvious? A used-car dealer will tell you that a car is a lovely little runner with one previous owner because they want you to buy the car, priests tell you that you must come to church so you can go to heaven, because otherwise they would be out of a job, and scientists tell us all that nonsense so we will be amazed at how clever they are and keep spending taxpayers' money on their research grants.

Alice: Now you are just being cynical; not everyone is out for themselves you know.

Thomas: And you are just being naïve; anyway, even supposing that scientists really believe their theories, can't you see that science is just the modern religion?

Alice: What do you mean?

Thomas: Well, if you were living five hundred years ago you would believe in angels and saints and the Garden of Eden; science has just replaced religion as the dominant belief system of the West. If you were living in a tribe in the jungle somewhere you would believe in whatever creation myths the elders of the tribe passed down to you, but you happen to be living here and now, so you believe what the experts in our tribe, who happen to be the scientists, tell us.

Alice: You can't compare religious dogma and myth with science.

Thomas: Why not?

Alice: Because scientists develop and test their beliefs according to proper methods rather than just accepting what they are told.

Thomas: Well you are right that they *claim* to have a method that ensures their theories are accurate but I don't believe it myself, otherwise they would all come to the same conclusions and we know that scientists are always arguing with each other, like about whether salt or sugar is really bad for you.

Alice: Well it takes time for theories to be proven but they will find out eventually.

Thomas: Your faith is astounding – and you claim that science and religion are totally different. The scientific method is a myth put about by scientists who want us to believe their claims. Look at all the drugs that have been tested by scientific methods and pronounced safe only to be withdrawn a few years later when people find out how dangerous they are.

Alice: Yes but what about all the successful drugs and the other amazing things science has done.

Thomas: Trial and error, that's the only scientific method there is, it's as simple as that. The rest is just propaganda.

Alice: I can't believe that; scientific theories, like the Big Bang theory, are proved by experiments and observations, that is why we ought to believe them and that is what makes them different from creation myths and religious beliefs.

Thomas: So you say but how can experiments and observations prove a theory to be true?

Alice: I suppose I don't really know.

Thomas: Well let me know when you've found out.

——o◯o——

In this dialogue, one of the characters challenges the other to explain why her beliefs, which are based on what she has been told by scientists, are any better supported than belief in angels and devils or the spirits and witchcraft of animistic religions. Of course, there are lots of things that each of us believe that we cannot justify directly ourselves; for example, I believe that large doses of arsenic are toxic to humans, but I have never even seen any arsenic as far as I am aware, and I have certainly never tested its effects. We all believe all kinds of things to be the case because we rely upon what others tell us directly or indirectly; whether or not we are justified depends upon whether or not they are justified. Most readers of this book probably believe that the Earth revolves around the Sun, that we as human beings evolved from animals that were more like apes, that water is made of twice as much hydrogen as oxygen, that diseases are often caused by viruses and other tiny organisms, and so on. If we believe these things it is because the experts in our tribe (the scientists) tell us them; in that way, the causes of our beliefs are of much the same kind as those of

13

someone who believes what the local witch-doctor tells them about, say, the cause of disease being the witchcraft of another person. We like to think that there is a difference between our beliefs and belief in witchcraft nonetheless; if there isn't then why do we spend so much money on modern drugs and treatments when a few sacrifices or spells would do just as well?

Our believer (Alice) thinks that the scientific method is what makes the difference, in that our beliefs are ultimately produced and proven by it, and that it has something to do with experiments and observation. In this chapter we will investigate the nature of the scientific method, if indeed there is one, beginning with the origins of modern science in the search for a new method of inquiry to replace reliance on the authority of the Church and the pronouncements of the ancients. Our goal will be to determine whether Alice, who believes in what science tells her, is entitled to her faith or whether the attitude of the sceptic, Thomas, is in fact the more reasonable one.

1.2   The scientific revolution

The crucial developments in the emergence of modern science in the western world took place during the late sixteenth and the seventeenth centuries. Within a relatively short space of time, not only was much of what had previously been taken for granted discredited and abandoned, but also a host of new theoretical developments in astronomy, physics, physiology and other sciences were established. The study of the motion of matter in collisions and under the influence of gravity (which is known as mechanics) was completely revolutionised and, beginning with the work of Galileo Galilei (1564–1642) in the early sixteen hundreds and culminating in the publication of Isaac Newton's (1642–1727) mathematical physics in 1687, this part of physics became a shining example of scientific achievement because of its spectacular success in making accurate and precise predictions of the behaviour of physical systems. There were equally great advances in other areas and powerful new technologies, such as the telescope and microscope, were developed.

This period in intellectual history is often called *the Scientific revolution* and embraces *the Copernican revolution*, which is the name given to the period during which the theory of the solar system and the wider cosmos, which had the Earth at the centre of everything (geocentrism), was replaced by the theory that the Earth revolved around the Sun (heliocentrism). From the philosophical point of view the most important development during the scientific revolution was the increasingly widespread break with the theories of Aristotle (384–322 BC). As new ideas were proposed, some thinkers began to search for a new method that could be guaranteed to bring knowledge. In the Introduction we found that for a belief to count as knowledge it must be justified, so if we want to have knowledge we might aim to follow a procedure when forming our beliefs that simultaneously provides us with a justification for them; the debate about what such a procedure might consist of, which happened during the scientific revolution, was the beginning of the modern debate about scientific method.

In medieval times, Aristotle's philosophy had been combined with the doctrines of Christianity to form a cosmology and philosophy of nature (often called *scholasticism*) that described everything from the motions of the planets to the behaviour of falling bodies on the Earth, the essentials of which were largely unquestioned by most western intellectuals. According to the Aristotelian view, the Earth and the heavens were completely different in their nature. The Earth and all things on and above it, up as far as the Moon, were held to be subject to change and decay and were imperfect; everything here was composed of a combination of the elements of earth, air, fire and water, and all natural motion on the Earth was fundamentally in a straight line, either straight up for fire and air, or straight down for water and earth. The heavens, on the other hand, were thought to be perfect and changeless; all the objects that filled them were supposed to be made up of a quite different substance, the fifth essence (or quintessence), and all motion was circular and continued forever.

Although not everyone in Europe prior to the scientific revolution was an Aristotelian, this was the dominant philosophical outlook, especially because of its incorporation within official Catholic doctrine. The break with Aristotelian philosophy began slowly and with great controversy, but by the end of the seventeenth century the radically non-Aristotelian theories of Galileo, Newton and others were widely accepted. Perhaps the most significant event in this process was the publication in 1543 of a theory of the motions of the planets by the astronomer Nicolaus Copernicus (1473–1543). In the Aristotelian picture, the Earth was at the centre of the universe and all the heavenly bodies, the Moon, the planets, the Sun and the stars revolved around the Earth following circular orbits. An astronomer and mathematician called Ptolemy of Alexandria (circa AD 150) systematically described these orbits mathematically. However, the planets' motions in the sky are difficult to reproduce in this way because sometimes they appear to go backwards for a while (this is called retrograde motion). Ptolemy found that to get the theory to agree at all well with observations, the motions of the planets had to be along circles that themselves revolved around the Earth, and this made the theory very complex and difficult to use (see Figure 1).

Copernicus retained the circular motions but placed the Sun rather than the Earth at the centre of the system, and then had the Earth rotating both about its own axis and around the Sun, and this considerably simplified matters mathematically. Subsequently, Copernicus' theory was improved by the work of Johannes Kepler (1571–1630), who treated the planets as having not circular but elliptical orbits, and it was the latter's theory of the motions of the planets that Newton elaborated with his gravitational force and which is still used today for most practical purposes.

One thing to note about the Copernican system is that it may seem to be counter to our experience in the sense that we do not feel the

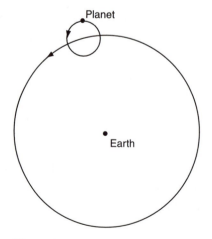

Figure 1

Earth to be moving when we stand still upon it, and moreover we observe the Sun to move over our heads during the day. This is an important example of how scientific theories seem to describe a *reality* distinct from the *appearance* of things. This distinction between appearance and reality is central to metaphysics because the latter seeks to describe things 'as they really are' rather than how they merely appear to be. When Copernicus' book was published, after his death, it included a preface by Andreas Osiander (1498–1552) (a friend of Copernicus who had helped prepare the book for publication) which declared that the motion of the Earth was a convenient assumption made by Copernicus but which need only be regarded as a mathematical fiction, rather than being taken literally as asserting that the Earth really was in orbit around the Sun. This is an early example of the philosophical thesis of **instrumentalism**, according to which scientific theories need not be believed to be true, but rather should be thought of as useful or convenient fictions. On the other hand, to be a realist about Copernicus' theory is to think that it should be taken literally and to believe that the Earth really does orbit the Sun. Realists, unlike instrumentalists, think that scientific theories can answer metaphysical questions. (We shall return to the **realism** versus instrumentalism debate later.)

The doctrine that the Earth is not at the centre of the universe and that it is, in fact, in motion around the Sun was in direct contradiction with Catholic doctrine and Osiander's preface did not prevent a controversy arising about Copernicus' theory. This controversy became quite fierce by the early years of the seventeenth century and, in 1616, Copernicus' book and all others that adopted the heliocentric hypothesis were placed on a list of books that Catholics were banned from teaching or even reading. It may be hard to appreciate why the Church was so worried about a theory in astronomy, but heliocentrism not only conflicted with the Aristotelian picture of the universe and rendered its explanations of motion inapplicable, it also conflicted with the traditional understanding of the Book of Genesis and the Fall of Adam and Eve, the relationship between the Earth and the Devil on the one hand and the Heavens and God on the other, and so on. The consequence of this was that if one were to adopt the Copernican theory, a great deal of what one took for granted was thrown into doubt – hence the need for a way of replacing the Aristotelian picture of the world with a set of beliefs that were equally comprehensive, but more up to date.

1.3   The 'new tool' of induction

The emergence of modern science required not just the contribution of those like Copernicus and Galileo who proposed new theories, but also the contribution of people who could describe and then advocate and propagate the new ways of thinking. In modern parlance, science needed to be marketed and sold to intellectuals who would otherwise have accepted the established Aristotelian thinking. Greatest among the propagandists of the emerging sciences was Francis Bacon (1561–1626), who explicitly proposed a method for the sciences to replace that of Aristotle. In his book *Novum Organum* of 1620 he set out this method in great detail and it still forms the core of what many people take the scientific method to be. Many of Bacon's contemporaries thought that the ancients had understood all there was to be known and that it was just a matter of recovering what had been lost. By contrast, Bacon was profoundly ambitious about what new things could be known and how such knowledge could be employed practically (he is often credited with originating the phrase 'knowledge is power').

Bacon's method is thoroughly egalitarian and collectivist in spirit: he believed that if it was followed by many ordinary people working together, rather than a few great minds, then as a social process it would lead to the production of useful and sure beliefs about the functioning of nature. When one bears in mind that nowadays a single paper in physics is routinely co-authored by tens of people, it is apparent that Bacon was prophetic, both in his vision of science as a systematic and collaborative effort involving the co-ordinated labour of many individuals to produce knowledge, and in his belief that the practical applications of science would enable people to control and manipulate natural phenomena to great effect. (On the other hand, one consequence of the growth of scientific knowledge has been that a great deal of training is now necessary before someone can become a researcher in, say, microbiology or theoretical physics.)

The translation of *Novum Organum* is New Tool, and Bacon

proposed his method as a replacement for the *Organum* of Aristotle, this being the contemporary name for the textbook that contained Aristotelian logic. Logic is the study of reasoning abstracted from what that reasoning is about. Hence, in logic the following two arguments are treated as if they were the same because their form or structure are equivalent despite the difference in their content:

(1) All human beings are mortal (PREMISE)
Socrates is a human being (PREMISE)
Therefore Socrates is mortal (CONCLUSION)

(2) All guard dogs are good philosophers
Fido is a guard dog
Therefore Fido is a good philosopher

The premises of the first argument are true and so is the conclusion, while the first premise of the second argument is probably false and so is the conclusion. What they have in common is that they exemplify the following structure:

All Xs are Y
A is X
Therefore A is Y

Such an argument is *valid*, which is to say if the premises are true then so must be the conclusion; in other words, if an argument is valid then it is *impossible* for the premises all to be true and the conclusion false.

An *invalid* argument is one in which the premises may all be true and the conclusion false, so for example, consider:

All Xs are Ys
A is Y
Therefore A is X

This argument is invalid as we can see if we have the following premises and conclusion:

All guard dogs are good philosophers
James is a good philosopher
Therefore James is a guard dog

Even if we suppose the first and second premises to be true, implausible as they may seem, it does not follow that James is a guard dog. (To reason in accordance with an invalid form of argument is to fall prey to a *logical fallacy*.) That this argument form is invalid is obvious when we consider the following argument that has the same structure but true premises and a false conclusion:

> All human beings are animals
> Bess is an animal
> Therefore Bess is a human being

Here we have an instance of the same form of argument where it is obviously possible for the premises to be true and the conclusion false (actually Bess is a dog) and hence it must be invalid. (Make sure you understand why this argument has the same form as the one immediately preceding it, and why both are invalid. It is important that validity has nothing to do with whether the premises or conclusion are actually true or false; it is a matter of how the premises and conclusion are related in form or structure. If a valid argument happens to have true premises it is said to be *sound*.)

Deductive logic is the study of valid arguments and Aristotelian logic is a type of deductive logic. The paradigm of deductive reasoning in science is Euclidean geometry. From a small number of premises (called axioms) it is possible to deduce an enormous number of conclusions (called theorems) about the properties of geometric figures. The good thing about deductive logic is that it is truth-preserving, which is to say that if you have a valid argument with true premises (such as argument (1)), then the conclusion will be true as well. The problem with deductive logic is that the conclusion of a deductively valid argument cannot say more than is implicit in the premises. In a sense, such arguments do not expand our knowledge because their conclusions merely reveal what their premises already state, although where the argument is complex we may find the conclusion surprising just because we hadn't noticed that it was already implicit in the premises, as with Pythagoras' theorem for example. Where the argument is simple, the fact that the conclusion says nothing new is obvious: if I already know that all humans are mortal, and that I am a human, I don't really learn anything from the conclusion that I am mortal, although I may find it strikes me with more force when it is made explicit.

The Aristotelian conception of knowledge (or scientia) restricts the domain of what is knowable to what is necessary and cannot be otherwise. Knowledge of some fact about the natural world, for example that flames go upwards but not downwards, consists of having a deductive argument that demonstrates the causal necessity of that fact from first principles; in this case, all things seek their natural place, the natural place of the element of fire is at the top of the terrestrial sphere, therefore flames near the surface of the Earth rise. In this view, geometry (in particular) and mathematics (in general) provide a model for knowledge of the natural world. Hence, the premises that one proceeds with have to concern the essence of the relevant entities. This knowledge of the essence of things, say that the natural place of fire is at the top of the terrestrial sphere, is presupposed by a demonstration, so the natural question is where does this knowledge of essences come from? The Aristotelian answer to this appeals to a kind of faculty of intellectual intuition that allows someone to perceive the causes of things directly, and among the causes that Aristotelian scientific inquiry aims to determine are the final causes of things, which is to say the ends towards which they are moving. Hence, Aristotelian science is concerned with **teleology**, which is the study of purposive behaviour.

The obvious objection to all this from the modern point of view is that there is little about the role of actual sensory experience in the acquisition of knowledge of how things work. If we want to know whether metals expand when heated we expect to go out and look at how metal actually behaves in various circumstances, rather than to try and deduce a conclusion from first principles. To the modern mind, science is immediately associated with experiments and the gathering of data about what actually happens in various circumstances and hence with a school of thought in epistemology called **empiricism**. Empiricists believe that knowledge can only be obtained through the use of the senses to find out about the world and not by the use of pure thought or reason; in other words, the way to arrive at justified beliefs about the world is to obtain evidence by making observations or gathering data. Aristotle's logic was deductive and, although he took great interest in empirical data and his knowledge of natural phenomena, especially zoology and botany, was vast, apparently he never carried out any experiments. Bacon proposed his 'inductive logic' to replace Aristotelian methods and gave a much more central role to experience and experiments.

Remember, as we saw in the discussion of Fido the guard dog, not all valid arguments are good ones. Another example of a valid but bad argument is the following:

> The Bible says that God exists
> The Bible is the word of God and therefore true
> Therefore God exists

This argument is deductively valid because it is not possible for the premises both to be true and the conclusion false, and indeed it may even have true premises, but it is not a good argument because it is circular; we only have a reason to believe that the second premise is true if the conclusion is true, and so a non-believer is unlikely to be persuaded by it. Similarly, perhaps not all invalid arguments are intuitively bad arguments. For example:

> Jimmy claims to be a philosopher
> I have no reason to believe he is lying
> Therefore Jimmy is a philosopher

This argument is invalid because it is possible for both premises to be true, but for the conclusion to be false, but it is nonetheless persuasive in ordinary circumstances. Validity is a formal property of arguments. Inductive reasoning, or **induction**, is the name given to various kinds of deductively invalid but allegedly good arguments. What distinguishes bad invalid arguments from good ones, if indeed there are any of the latter? Bacon claims to have an answer to this question that vastly improves on Aristotle's answer. A large part of what Bacon advocates is negative in the sense that it amounts to a way of avoiding falling into error when making judgements rather than offering a way of gaining new judgements. This negative side to the scientific method is recognisable in science today when people insist that to be a scientist one must be sceptical and prepared to break with received wisdom, and also not leap to conclusions early in the process of investigation of some phenomenon. Bacon called the things that could get in the way of right inductive reasoning the *Idols of the Mind* (which are analogous to fallacies of reasoning in deductive logic).

The first of these are the *Idols of the Tribe*, which refers to the tendency of all human beings to perceive more order and regularity in nature than there is in reality, for example, the long-standing view mentioned above that all heavenly bodies move in perfect circles, and to see things in terms of our preconceptions and ignore what doesn't fit in with them. The *Idols of the Cave* are individual weaknesses in reasoning due to particular personalities and likes and dislikes; someone may, for example, be either conservative or radical in temperament and this may prejudice them in their view of some subject matter. The *Idols of the Marketplace* are the confusions engendered by our received language and terminology, which may be inappropriate yet which condition our thinking; so, for example, we may be led into error by our using the same word for the metal lead and for that part of a pencil that makes a mark on paper. Finally, the *Idols of the Theatre* are the philosophical systems that incorporate mistaken methods, such as Aristotle's, for acquiring knowledge.

So much for the negative aspects of Bacon's philosophy, but what of the positive proposals for how to acquire knowledge of the workings of the natural world? His method begins with the making of observations that are free from the malign influence of the first three Idols. The idea is to reach the truth by gathering a mass of information about particular states of affairs and building from them step by step to reach a general conclusion. This process is what Bacon called the composition of a Natural and Experimental History. Experiments are important because if we simply observe what happens around us we are limited in the data we can gather; when we perform an experiment we control the conditions of observation as far as is possible and manipulate the conditions of the experiment to see what happens in circumstances that may never happen otherwise. Experiments allow us to ask 'what would happen if . . .?'. Bacon says that by carrying out experiments we are able to 'torture nature for her secrets'. (Some feminist philosophers have emphasised that the conception of science as the masculine torture of feminine nature was very common in the scientific revolution and have argued that the science that we have today has inherited this gender bias.)

Experiments are supposed to be repeatable if at all possible, so that others can check the results obtained if they wish. Similarly, scientists prefer the results of experiments to be recorded by instruments that measure quantities according to standard definitions and scales so that the perception of the individual performing the experiment does not affect the way the outcome is reported to others. Bacon stressed the role of instruments to eliminate, as far as possible, the unreliable senses from scientific data gathering. In this way the scientific method of gathering data that will count as evidence for or against some view or other is supposed to ensure objectivity or impartiality. It seems obvious to the modern mind that science is all to do with experiments, but prior to the scientific revolution experiments were mainly associated with the practices of alchemists, and experiments played almost no role in Aristotle's methods.

Having gathered data from naturally occurring examples of the phenomenon we are interested in, as well as those produced by the ingenious manipulation of experimental design, we must then put the data in tables of various kinds. This process is best illustrated with Bacon's own example of the investigation of the phenomenon of heat. The first table to be drawn up is that of Essence and Presence, which consists of a list of all the things of which heat is a feature, for example, the Sun at noon, lava, fire, boiling liquid, things that have been vigorously rubbed and so on. The next table is that of Deviation and Absence by Proximity, which includes things that are as close to the above phenomena as possible but which differ by not involving heat; so, for example, the full Moon, rock, air, water that is cold, and so on. One big problem with the little that Aristotle did say about induction, as far as Bacon was concerned, was that it seemed to sanction the inference from particular instances straight to a generalisation without the mediation of so-called middle axioms. For Bacon the advantage of his inductive method was that it would avoid this problem by searching for negative instances and not just positive ones. There follows a table of Degrees or Comparisons in which the phenomena in which heat features are quantified and ranked according to the amount of heat they involve.

Having drawn up all these tables, the final stage of Bacon's method is the Induction itself. This involves studying all the information displayed in the tables and finding something that is present in all instances of the phenomenon in question, and absent when the phenomenon is absent, and furthermore, which increases and decreases in amount in proportion with the increases and decrease of the phenomenon. The thing that satisfies these conditions is to be found by elimination and not by merely guessing. Something like the method of elimination is used by people all the time, for example, when trying to find the source of a fault with an electrical appliance such as a hi-fi system. First, one might try another appliance in the same socket; if it works then the socket is not to blame so one might next change the fuse, if the system still does not work the fuse is not to blame so one might check the connections in the plug, then one might test the amplifier, and so on. In the case of heat Bacon decides that heat is a special case of motion, in particular the 'expansive motion of parts' of a thing. This accords remarkably well with the modern understanding of heat (which was not developed until the mid-nineteenth century), known as the kinetic theory of heat according to which heat consists of molecular motion, and the faster the average velocity of the molecules in some substance then the hotter it will be.

According to Bacon, the **form** of expansive motion of parts is what underlies the phenomenon of heat as it is observed. Bacon thought that, following his method, one could discover the forms, which, although not directly observable, produce the phenomena that we can perceive with the senses. Once knowledge of the true forms of things was obtained then nature could be manipulated and controlled for the benefit of people. Bacon suggested that the kind of power over nature that was claimed by magicians in the Renaissance could be achieved through scientific methods. If we consider the development of science and technology since Bacon's time it certainly seems that technology has accomplished feats that surpass the wildest boasts of magicians: who would have believed a magus who claimed to be able to travel to the Moon or to the depths of the oceans; who would have imagined synthesising the materials out of which computers are made, or the transmission of images by photograph, film and television?

When Bacon says that science ought to discover the forms of things, he means, as in the case of heat, the concrete and immediate physical causes of them, and not the final causes that Aristotelians aimed to find by direct intuition, such as the cause of the motion of a dropped stone towards the Earth being the fact that the 'natural place' of the element of which the stone is composed is at the centre of the Earth. Such explanations seemed vacuous to Bacon, as with the notorious claim that opium sends people to sleep because it possesses a dormative virtue. The abandonment of the search for final causes was one of the main consequences of the scientific revolution. By the eighteenth century, the French writer Voltaire (1694–1778) in his play *Candide* was ridiculing the Aristotelian model of explanation; the character Doctor Pangloss explains the shape of the nose of human beings in terms of its function in holding a pair of glasses on the face. Bacon explicitly urged that teleological reasoning be confined to the explanation of human affairs where it is legitimate since people are agents who act so as to bring about their goals. One characteristic of natural science since Bacon is that explanations are required to refer only to the immediate physical causes of things and the laws of nature that govern them. (Whether or not this requirement is satisfied is a controversial issue, especially because evolutionary biology has reintroduced talk of functions and design into science. However, it is often claimed that such talk is only legitimate because it is, in principle, eliminable or reducible to a series of proper causal explanations. We shall return to this issue in Chapter 7.)

So the 'forms' of Bacon are the immediate causes or the general principles or laws that govern phenomena in the material world. However, Bacon's account of scientific theorising leaves us with a problem to which we shall return throughout this book, namely how exactly do we come to conceive of the forms of things given that they are not observable? In the case of heat we may be relatively happy with Bacon's induction, but motion is a feature of the observable world too and not confined to the hidden forms of things. When it comes to something like radioactivity, which has no observable counterpart, how could we ever induce its presence from tables like Bacon's? Baconian induction is meant to be a purely mechanical procedure but there will be many cases where no single account of the form of some phenomenon presents itself and where different scientists suggest different forms for the same phenomenon; an example is the debate about the nature of light which concerned two theories, a wave theory and a particle theory.

Bacon does offer us something else that may help with this problem, which is his notion of a 'pejorative instance' (although this is the subject of great controversy, as we shall see). He argues that when we have two rival theories that offer different accounts of the form of something then we should try and design an experiment that could result in two different outcomes where one is predicted by one theory and the other by the other theory so that, if we perform the experiment and observe the actual outcome, we can choose between them. (The great seventeenth century scientist Robert Hooke (1635–1703) called such experiments 'crucial experiments'.) An example Bacon suggests is an experiment to see if gravity is really caused by the force of attraction produced by large bodies like the planets and the Sun; if this is really so then a clock that works by the gravitational motion of a pendulum ought to behave differently if it were placed up a church tower, or down a mine (further from, or closer to, the centre of the Earth respectively), hence, performing this experiment ought to allow us to tell whether the attractive hypothesis is correct. (In fact, the gravitational attraction of the Earth is stronger down a mine-shaft than up a tower, but the difference is very small and hence very hard to detect.)

This is an important idea because it implies that experiments in science will not be a simple matter of going out and gathering data but rather will involve the designing of experiments with the testing of different theories already in mind. This may seem to undermine Bacon's claim that we should record our natural and experimental history of the phenomenon we are studying without being influenced by our preconceptions (and so avoid the Idols of the Theatre), however, Bacon would argue that the need for pejorative instances will only arise once we have carried out our initial investigations and ended up with more than one candidate for the form of the phenomenon.

1.4 (Naïve) inductivism

We can abstract Bacon's method and arrive at a simple account of the scientific method. The method of Bacon rested on two pillars, *observation* and *induction*. Observation is supposed to be undertaken without prejudice or preconception, and we are to record the results of the data of sensory experience, what we can see, hear, and smell, whether of the world as we find it, or of the special circumstances of our experiments. The results of observation are expressed in what are called *observation statements*. Once we have made a whole host of observations these are to be used as the basis for scientific laws and theories. Many scientific laws are of the form of what are called *universal generalisations*; these are statements that generalise about the properties of all things of a certain kind. So, for example, 'all metals conduct electricity' is a universal generalisation about metals, 'all birds lay eggs' is a universal generalisation about birds, and so on. These are simple examples but, of course, scientific theories are often much more complicated and the generalisations and laws often take the form of mathematical equations relating different quantities. Some well known examples include:

- *Boyle's law*, which states that for a fixed mass of a gas at constant temperature, the product of pressure and volume is constant.
- Newton's *law of universal gravitation*, which states that the gravitational force, F, between two bodies with masses m_1, m_2, and separated by distance r, is given by: $F = m_1 m_2 G/r^2$ (where G is the gravitational constant).
- The *law of reflection*, which states that the angle at which a beam of light strikes a mirror is equal to the angle at which it is reflected.

Induction in the broadest sense is just any form of reasoning that is not deductive, but in the narrower sense that Bacon uses it, it is the form of reasoning where we generalise from a whole collection of particular instances to a general conclusion. The simplest form of induction is *enumerative induction*, which is where we simply observe that some large number of instances of some phenomenon has some characteristic (say some salt being put in a pot of water dissolves), and then infer that the phenomenon always has that property (whenever salt is put in a pot of water it will dissolve). Sometimes scientific reasoning is like this, for example, many of the drug and other medical treatments that are used today are based on trial and error. Aspirin was used to relieve headaches a long time before there were any detailed explanations available of how it worked, simply because it had been observed on many occasions that headaches ceased following the taking of the drug.

The question that we must now ask is: 'when is it legitimate to infer a universal generalisation from a collection of observation statements?', for example, when can we infer that 'all animals with hearts have livers' on the basis of the observation of many instances of animals having hearts having livers as well. The answer according to naïve **inductivism** is *when a large number of observations of Xs under a wide variety of conditions have been made, and when all Xs have been found to possess property Y, and when no instance has been found to contradict the universal generalisation 'all Xs possess property Y'*. So, for example, we need to observe many kinds of animals in all parts of the Earth, and we need to look out for any instance that contradicts our generalisation. If we carry out a lot of observations and all support the law while none refute it, then we are entitled to infer the generalisation.

This accords with our common sense; someone who concluded that all philosophers are neurotic, having observed only a handful of philosophers in Bristol to be neurotic, would be considered quite unreasonable. Similarly, someone who drew such an inference having observed one perfectly stable and balanced philosopher would be considered unreasonable no matter how many other philosophers they had observed showing signs of neurosis. However, if someone claimed to believe that all philosophers are neurotic and when questioned it turned out they had observed philosophers both young and old, of both sexes and in various parts of the world over many years and they had all been neurotic to varying degrees and not one had no trace of neurosis, we would think their conclusion quite reasonable in the circumstances.

What we have just been discussing is known as a *Principle of Induction*; it is a principle of reasoning that sanctions inference from the observation of particular instances to a generalisation that embraces them all and more. We must take care to observe the world carefully and without preconception, and to satisfy the conditions expressed in the principle, but if we do this then, according to the naïve inductivist, we are following the scientific method and our resulting beliefs will be justified. Once we have inductively inferred our generalisation in accordance with the scientific method, then it assumes the status of a law or theory and we can use deduction to deduce consequences of the law that will be predictions or explanations.

It's time we caught up with the discussion with which this chapter began:

———·◦◯◦·———

Alice: . . . and so the scientific method consists in the unbiased accumulation of observations and inductive inference from them to generalisations about phenomena.

Thomas: But even if I buy that for claims about metals conducting electricity and the like, which I don't, I still don't see how induction explains how we know about atoms and all that stuff you were going on about before.

Alice: I guess it's to do with Bacon's idea about crucial experiments; someone says that there are atoms and someone else works out how to do an experiment that ought to go one way if there are atoms and another way if there are not.

Thomas: Well anyway, let's forget about atoms for now and just concentrate on your principle of induction and Bacon's idea about observation without prejudice or preconception. I can already think of problems with both of these; for one thing, how do you know that your principle of induction is true, and for another, how would you know what to start observing unless you already had the idea of metals and electricity? Observation without any bias whatsoever is impossible, and you haven't explained to me why I should believe in induction. I still reckon that science is just witchcraft in a white coat.

———·◦◯◦·———

Further reading

For an excellent account of the scientific revolution see Steven Shapin *The Scientific Revolution* (Chicago University Press, 1996). Another introductory book is I. Bernard Cohen, *The Birth of a New Physics* (Pelican, 1987). On Francis Bacon see Chapter 3 of Barry Gower, *Scientific Method: An Historical and Philosophical Introduction* (Routledge, 1997), Chapter 2 of Roger Woolhouse, *The Empiricists* (Oxford University Press, 1988), Peter Urbach, *Francis Bacon's Philosophy of Science: An Account and a Reappraisal* (Open Court, 1987), and also the references to Bacon's works in the bibliography.

2

—∘⊙∘—

The problem of induction and other problems with inductivism

According to the account of scientific method that was introduced in the previous chapter (naïve inductivism), scientific knowledge derives its justification by being based on generalisation from experience. Observations made in a variety of circumstances are to be recorded impartially and then induction is used to arrive at a general law. This is an attractive view, not least because it agrees with what many scientists have claimed about their own practice. It also explains the alleged objectivity of scientific knowledge by reference to the open-mindedness of scientists when they make observations, and it keeps scientific knowledge firmly rooted in experience. I hope it is a reasonably familiar conception of how science works and how scientific knowledge acquires its justification.

We need to distinguish two questions in order to evaluate inductivism as a theory of scientific methodology:

(1) Does inductivism seem to be the method that has actually been followed by particular individuals in the history of science?
(2) Would the inductive method produce knowledge if we did use it?

The first question obviously calls for some empirical inquiry; to answer it we need to gather information from artefacts, journals, letters, testimony and so on. The second question is characteristically philosophical and concerns not our actual beliefs but whether the inductive method will confer justification on beliefs that are produced using it. We will return to question (1) later, while in the next section we will consider whether or not induction is justified.

2.1 The problem of induction

The classic discussion of the problem of induction is in *An Enquiry Concerning Human Understanding* by David Hume (1711–1776). Hume relates induction to the nature of causation and the laws of nature, and his influence on the development of western philosophy in general, and philosophy of science in particular, has been profound. To understand Hume's arguments about scientific knowledge it will be helpful to have a basic grasp of his general epistemology and theory of 'ideas'.

Hume makes a distinction between two types of proposition, namely those that concern *relations of ideas* and those that concern *matters of fact*. The former are propositions whose content is confined to our concepts or ideas, such as a horse is an animal, bachelors are unmarried, and checkmate is the end of a game of chess. (Hume also included mathematics in this category, so triangles have angles totalling 180° is another example.) Propositions concerning matters of fact are those that go beyond the nature of our concepts and tell us something informative about how the actual world is. So, for example, snow is white, Paris is the capital of France, all metals expand when heated, and the battle of Hastings was in 1066 are all propositions that concern matters of fact. Of course, these propositions are all true (as far as I know), but the distinction between relations of ideas and matters of fact applies equally to propositions that are false, so for example, a whale is a fish is a false proposition concerning relations among our ideas, and Plato died in 399 BC is a false proposition concerning a matter of fact.

According to Hume, any true proposition about the relations among our ideas is provable by deduction, because its **negation** will imply a contradiction. Those who have studied mathematics or logic will be familiar with the method of *reductio ad absurdum*. Essentially, the idea is that some proposition, say that there are an infinite number of prime numbers, can be proved if you can show that the negation of it is inconsistent with other things you already know. Such a proof would begin with the assumption that there is a biggest prime number. This is then used in conjunction with other assumed facts about numbers (in particular, about the existence of prime factors) to derive a contradiction. (Not all proofs have this form on the surface but the definition of a logically necessary truth is that its negation is a contradiction.) In everyday life, something similar to this method is also sometimes employed when people try to show that an absurd or known to be false consequence follows from some proposition under discussion.

On the other hand, Hume argued that knowledge of matters of fact could only be derived from the senses because the ideas involved are logically unrelated and hence the propositions are not deductively provable. Take the proposition that Everest is the tallest mountain on Earth. The concepts involved – mountain, tallest, Earth, and that of some specific mountain in the Himalayas – have no logical relation to each other that determines the truth of the proposition, and there is no contradiction in supposing that some other mountain is the tallest. Hence, it is not possible to find out if the proposition is true merely by reasoning; only by using the senses can the status of such propositions be investigated. (Hume, who was Scottish, is a central figure in the philosophical tradition known as British empiricism, which also includes the English John Locke (1632–1704) and the Irish George Berkeley (1685–1753).) All these thinkers shared the belief that there are no innate concepts and that all our knowledge of the world is derived from, and justified by, our sensory perceptions, hence they all deny that any *a priori* knowledge of matters of fact is possible.

Hume was also very sceptical about metaphysical or theological speculation. Now, many people, including some philosophers, think that philosophy is often concerned with concepts so abstract and distanced from everyday life that they have no bearing on anything one could measure or experience, and that because of this they are more or less meaningless. Some people would also argue that thinking in this manner is a waste of time. Hume agreed and suggested that if one takes some book, or other text, and it contains neither 'abstract reasoning concerning quantity or number', nor 'experimental reasoning concerning matter of fact and existence', then it should be burned since it is merely 'sophistry and illusion'. This dichotomy is known as *Hume's fork*. (I leave it as an exercise for the reader to decide what ought to be done with the present volume.)

Hume's distinction between matters of fact and relations of ideas roughly corresponds to Immanuel Kant's (1724–1804) distinction between synthetic and **analytic** truths. Kant was inspired by Hume and made the latter distinction a central part of his (critical) philosophy. In the hands of a group of philosophers of science, called the *logical positivists*, in the early twentieth century, it became a way of distinguishing form from content in formal mathematical and logical languages that were used to represent scientific theories. They thought that they could separate the empirical content of theories, the synthetic part, from the theoretical and analytic part. The positivists argued that a factual statement was not meaningful if it said nothing about any past, present or future observations, in other words if it has no empirical content. This gives us a way of deciding whether someone is talking nonsense or not; we check to see if what he or she is saying has any implications for what we can observe. Positivism, which will often come up again (see especially 5.3), was very influential among philosophers and scientists for a while, and still has adherents. Many people sympathise with the idea that scientific and philosophical theories should have a definite connection to what can somehow be observed, and perhaps also measured, recorded and ultimately given a theoretical description in terms of laws and causes.

Now, it is plausible to argue that some of our knowledge of matters of fact is directly based on experience. That it is windy, cloudy and cold outside, that the light is on and the tea luke-warm, all this I seem to know by my present sensory experience. Another class of the things I know are those that I learned by the same means in the past; such knowledge is based on my memory of my perceptions. What of my beliefs about things I have not myself observed? I certainly have many such beliefs, for example, I believe that the Sun will rise tomorrow, that Everest is the tallest mountain, that my friend is currently in Scotland, and so on. These are all matters of fact because, in each case, the negation of the proposition is not a contradiction and so we cannot deductively prove them to be true. How can we *know* such things, if indeed we can?

Hume claimed that all reasoning that goes beyond past and present experiences is based on cause and effect. Suppose that you play pool a lot; it doesn't take long to notice that if you hit the white ball off centre it will impart a particular kind of spin to the next ball it hits. This is a useful generalisation about the behaviour of pool balls. You infer that hitting the ball off centre *causes* it to spin and that you can reliably predict the behaviour of the balls in future on this basis, provided of course you can hit them right. Similarly, we observe that when the Sun is out, the Earth and the objects on its surface become warmer and we infer that this pattern of behaviour will continue in the future and that the Sun causes the objects to heat up. Hume pointed out that there is nothing logically inconsistent in a pool ball suddenly spinning the opposite way or not at all, nor is there any contradiction in supposing that the Sun might cool down the Earth. The only way we connect these ideas is by supposing that there is some causal connection between them.

Of course, many of our beliefs depend upon the testimony of others, whether in the form of spoken accounts, books, newspapers, or whatever. In such cases we believe in a causal relation between what has happened or is the case, and what the person experiences and then communicates. Once again it is a causal relation that connects ideas that have no logical relation. This is the basis of induction according to Hume, and so if we want to understand our knowledge of matters of fact we need to consider our knowledge of the relation of cause and effect. Hume argues that we can only obtain our knowledge of cause and effect by experience because there is no contradiction in supposing that some particular causal relation does not hold, and so this knowledge is of a matter of fact that could be otherwise. We cannot tell that fire will burn us or that gunpowder will explode without trying it out because there is no contradiction in supposing that, for example, the next fire we test will not burn but freeze a hand placed in it. (Of course we may be told about causal relations, but then the source of our information is ultimately still someone's experience.)

What more can we say about this relation of cause and effect? Hume argues that, just as it is only by experience that we can find out about particular causal relations, and hence make inductive inferences about the future behaviour of things in the world, so it is only by examining our experience of the relation of cause and effect that we can understand its nature, and hence see whether it is fit to offer a justification for our inductive practices. When we examine our experience of causal relations, Hume argues that it is apparent that our knowledge of cause and effect is the result of extrapolating from past experience of how the world has behaved to how it will behave in future. For example, because the experience of eating bread has always been followed in the past by the experience of feeling nourished, I suppose that bread nourishes in general and hence that the next piece of bread I eat will be nourishing. Fundamentally then, for Hume, causation is a matter of what is known as *constant conjunction*; A causes B means A is constantly conjoined in our experience with B: 'I have found that such an object has always been attended with such an effect, and I foresee, that other objects, which are, in appearance similar, will be attended with similar effects' (Hume 1963: 34–35). But of course we have not yet experienced the future behaviour of the objects in question and so belief in a particular relation of cause and effect relies upon the belief that the future will resemble the past. (This is a crucial point to which we shall return below.)

Hume further examines the concept of causality and finds that an important feature of it is that of *contiguity*, which is the relation of being connected in space and time. Often, when a causal connection is postulated between events, the events are either close in space and time or connected by a chain of causes and effects, each member of which is close in space and time to the next. So, for example, there is a causal relation between someone typing words into a computer and someone else reading words on a page, because there is an intermediate chain of contiguous causes and effects, however long and complicated. However, Hume does not say that this is always the case where there is a postulated causal connection.

Another characteristic of causal relations is that causes usually precede effects in time. Whether this is always so is not immediately obvious, because sometimes it seems that causes and effects can be simultaneous, as when we say that the heavy oak beam is the cause of the roof staying up. Furthermore, some philosophers hold that 'backwards causation' where a cause brings about an effect in the past is possible. In any case, Hume has identified the following features that usually pertain to the relation A causes B:

(1) Events of type A precede events of type B in time.
(2) Events of type A are constantly conjoined in our experience with events of type B.
(3) Events of type A are spatio-temporally contiguous with events of type B.

(4) Events of type A lead to the *expectation* that events of type B will follow.

This is called the Humean analysis of causation, but is that all there is to causal relations? Consider the following example; a pool ball X strikes another Y, and Y moves off at speed. We say that X causes Y to move, but what does this mean? We are inclined to say things like the following; X made Y move, X produced the movement in Y, Y had to move because X hit it, and so on. Hume is well aware that many philosophers have held the view that X causes Y means that there is some sort of *necessary connection* between X happening and Y happening, but he argues that this notion is not one that we really understand. His empiricism led him to argue that since we have no experience of a necessary connection over and above our experience of constant conjunction, we have no reason to believe that there is anything corresponding to the concept of a necessary connection in nature. All we ever see are events conjoined; we never see the alleged connection between them, but over time we see the same kinds of events followed by similar effects and so we get into the habit of expecting this to continue in future.

In a form of argument we will return to later he argues as follows. Consider two theories about causation: according to the first, a causal relation consists of nothing more than the Humean analysis above reveals; according to the second there is all that but also some kind of necessary connection (call this the *necessitarian* view). Hume points out that there is nothing that can be found in our experience that will tell in favour of either one of these hypotheses over the other. These are two different hypotheses that agree about everything we can observe; yet one of them posits the existence of something that the other does not. Hence, Hume argues, we should adopt the Humean analysis because it does without metaphysical complications. Implicit in this argument is an appeal to the principle called 'Occam's razor' according to which, whenever we have two competing hypotheses, then if all other considerations are equal, the simpler of the two is to be preferred. Hume's empiricism means that he thinks that, because the two hypotheses entail exactly the same thing with respect to what we are able to observe, then all other considerations that are worth worrying about are indeed equal.

So, although our inductive reasoning is founded on reasoning about cause and effect, this is no foundation at all since it is always possible that a causal relation will be different in the future. Hume argues that the only justification we have for such beliefs as that the Sun will rise tomorrow, or that pool balls will continue to behave as they do, is that they have always been true up to now, and this isn't really any justification at all. Of course, we may appeal to the conservation of momentum and the laws of mechanics to explain why X caused Y to move. Similarly, we can now appeal to proofs of the stability of the solar system and predictions of the lifetime of the Sun to justify our belief that the Sun will rise tomorrow. However, Hume would say that the causal links and laws we are appealing to are just more correlations and regularities.

Fundamentally, Hume's problem with induction is that the conclusion of an inductive argument could always be false no matter how many observations we have made. Indeed, there are notable cases where huge numbers of observations have been taken to support a particular generalisation and it has subsequently been found to be wrong, as in the famous case of the generalisation *all swans are white* which was believed by Europeans on the basis of many observations until they visited Australia and found black swans. As Bertrand Russell (1872–1970) famously argued in the *Problems of Philosophy*, sometimes inductive reasoning may be no more sophisticated than that of a turkey who believes that it will be fed every day because it has been fed every day of its life so far, until one day it is not fed but eaten. The worrying thought is that our belief that the Sun will rise tomorrow may be of this nature.

Of course, we are capable of being more discriminating. Many of our beliefs seem to be based on something like the principle of induction that we discussed at the end of the previous chapter, which allows the inference from particular observations to a generalisation when there are many observations made under a wide variety of circumstances, none of which contradict the generalisation but all of which are instances of it. Yet, such a principle also expresses a tacit commitment to the uniformity of natural phenomena in space and time. But why should the future resemble the past or the laws of nature be the same in different places? Hume points out that the proposition that the future will not be like the past is not contradictory.

Of course, in the past we have observed patterns and believed that they will continue to hold in the future and we have been right. But for Hume this is just to restate the problem, for the fact that in the past the future has been like the past doesn't mean that, in the future, the future will be like the past. In other words, our past experience can only justify our beliefs about the future if we have independent grounds for believing that the future will be like the past, and we do not have such grounds.

Similarly, we might try and defend induction with an inductive argument along the lines of the following; induction has worked on a large number of occasions under a variety of conditions, therefore induction works in general. But Hume argues that this is viciously circular: it is inductive arguments whose justification is in doubt, therefore it is illegitimate to use an inductive argument to support induction, to do so would be like trying to persuade someone that what you have just said is true by informing them that you always tell the truth; if they already doubt what you have said then they already doubt that you always tell the truth and simply asserting that you do will not move them. By definition, in inductive arguments, it is possible the premises may all be true and the conclusion nonetheless false. So any defence of induction must either appeal to a principle of induction or presuppose the justification of inductive inference. Hence, Hume thought all justifications of induction are circular.

Note that, although we have taken inductive reasoning to be that which proceeds from past experience to some generalisation about the future behaviour of things, it is really the extrapolation from the observed to the unobserved that is at issue. Hume thinks that the same problem arises even if we infer not a generalisation but just some particular prediction, like that the Sun will rise tomorrow or that the next piece of bread I eat will be nourishing.

Of course, in order to survive we have to act in various ways and so we have no choice but to assume that the next piece of fresh bread we eat will be nourishing, that the Sun will rise tomorrow, and that in numerous other ways the future will be like the past. Hume does not think his scepticism seriously threatens what we actually believe and how we will behave. However, he also thinks that we will continue to make inductive inferences because of our psychological disposition to do so, rather than because they are rational or justified. It is our passions, our desires, and our animal drives that compel us to go beyond what reason sanctions and believe in the uniformity of nature and the relation of cause and effect.

To summarise, Hume observes that our inductive practices are founded on the relation of cause and effect, but when he analyses this relation he finds that all that it is, from an empiricist point of view, is the constant conjunction of events, in other words, the objective content of a posited causal relation is always merely that some regularity or pattern in the behaviour of things holds. Since the original problem is that of justifying the extrapolation from some past regularity to the future behaviour of things appealing to the relation of cause and effect is to no avail. Since it is logically possible that any regularity will fail to hold in the future, the only basis we have for inductive inference is the belief that the future will resemble the past. But that the future will resemble the past is something that is only justified by past experience, which is to say, by induction, and the justification of induction is precisely what is in question. Hence, we have no justification for our inductive practices and they are the product of animal instinct and habit rather than reason. If Hume is right, then it seems all our supposed scientific knowledge is entirely without a rational foundation.

2.2 Solutions and dissolutions of the problem of induction

Hume accepts that scepticism cannot be defeated but also that we have to get on with our lives. However, he argues that what is sometimes today called inductive reasoning, inductive inference or ampliative inference, is not really reasoning at all, but rather merely a habit or a psychological tendency to form beliefs about what has not yet been observed on the basis of what has already been observed. He is quite sure that, despite learning of the problem of induction, people will continue to employ induction in science and everyday life, indeed he thinks that we cannot help but do so in order to be able to live our lives, but he does not think this behaviour can be justified on rational grounds. Because of the way he tries to resolve philosophical problems by appealing to natural facts about human beings and their physiological and psychological make-up, Hume is an important figure in a philosophical tradition, called naturalism, that is particularly prominent in contemporary philosophy, although nowadays naturalists are not usually sceptics like Hume (recall from the Introduction that naturalists think that philosophy is continuous with empirical inquiry in science).

Most philosophers have not been satisfied with his sceptical naturalism and various strategies have been adopted to solve or dissolve the problem of induction. Note that some philosophers have employed more than one of the following.

(1) *Induction is rational by definition*

This response comes in crude and sophisticated versions; the crude version is as follows: in everyday life – in other words outside of academic philosophy – people do not use the term 'rational' to apply only to deductively valid inferences, indeed they often describe inductive inferences as rational. For example, consider three ways of making inferences about the fortunes of a football team based on past experience: if we are following the first method we predict the results of the next match by reading tea leaves; if we are following the second method we look at how the team did in their last few matches and then infer that they will do well next time if they did badly last time and vice versa; if we are following the third method we will again look at how the team did in their last few matches but then infer that they will do well next time if they did well last time and vice versa. Obviously the latter method is the one that everyone would say was the rational method, but this method is just the one that assumes that the future will be like that past and that nature is uniform. Indeed, most people would say that, in general, it is rational to base beliefs about the future on knowledge about the past. Hence, it is part of what everyone means by 'rational' that induction is rational.

This mode of philosophical argument was once very fashionable, but it is not sufficient to dispel philosophical worries about induction because when we ordinarily use a term like rational we are taking it to have some *normative* (or *prescriptive*) as well as descriptive content. In other words, we suppose that reasoning is rational because it conforms to some sort of standard and that it is the sort of reasoning that will tend to lead us to truth and away from falsity. Merely being called 'rational' is not enough to make a mode of reasoning justified, for it does not establish that the reasoning in question has the other properties that we take rational reasoning to have.

The second version of this response is more subtle. Instead of arguing that induction is rational because everyone uses the word 'rational' in a way that applies to it, we can argue that we are more certain of the general rationality of induction than we are of the validity of Hume's argument against it. In other words, we can treat Hume's argument like a paradox that leads to a conclusion that must be false (that induction is always irrational), and hence conclude that one or more of its premises must be false (although we may not be able to identify which one). This is, in fact, how most philosophers regard Hume's argument; they do not take it to show that induction is always irrational but rather to show that we do not know how to justify it. Adopting this strategy commits us to the task of working out exactly where the flaw is in Hume's argument, and also to giving some positive account of induction to replace Hume's negative one, but the point is that we may argue that there must be some such flaw even when we have no idea what it is. (Some philosophers argue that, in fact, this is the position that Hume himself held although most philosophers have taken him to be a sceptic who thinks that induction is unreasonable.)

(2)   Hume is asking for a deductive defence of induction, which is unreasonable

Some philosophers have accused Hume of demanding a deductive defence of induction. They argue that Hume assumes, without any argument, that deduction is the only possible source of justification for all beliefs other than those we directly experience or remember. Initially this claim is attractive, after all Hume doesn't say much about what inductive reasoning is like, other than it is not deductive, and he does seem to argue that induction is unjustified because of the fact that, in an inductive argument, it is possible that the premises are all true and the conclusion nonetheless false, which is just to say that the argument is deductively invalid. So it may look as if he is arguing that beliefs reached by inductive inference are unjustified just because the inference is non-deductive.

However, it is clear that Hume has more in mind than this because he diagnoses inductive inferences as all depending on the principle that nature is uniform. It is the fact that we have no independent reason to believe this principle that motivates scepticism about induction, in other words, because we have no reason to believe that nature is uniform in the sense that the future will resemble the past, then we have no reason to believe the conclusion of an inductive argument. This response is therefore not sufficient to dispel Hume's inductive scepticism.

(3) Induction is justified by the theory of probability

Many philosophers have tried to solve the problem of induction by appealing to the mathematical theory of probability. Perhaps the most detailed and sustained attempts of this kind were by Rudolf Carnap (1891–1970) and Hans Reichenbach (1891–1953), two of the greatest philosophers of science of the twentieth century. They tried to construct an *a priori* theory of inductive logic that would allow the calculation of the degree to which any particular hypothesis is confirmed. The problem with this strategy is that the application of technical results in mathematics to our knowledge of the world is impossible unless we make some substantial assumptions about how the world behaves, and such assumptions can never be justified on purely logical or mathematical grounds. Hence, we will still need to supplement our appeal to probability theory with some principle that assures us that it is applicable to the world (see the next strategy), and the problem will then be pushed back to the question of what justifies our belief that such a principle will hold in the future.

(4) Induction is justified by a principle of induction or of the uniformity of nature

One response to the problem of induction, which takes various forms, is to adopt some principle and insert it as a premise into inductive arguments to render them deductively valid. Suppose, for example, that we have often observed that sodium burns with an orange flame when heated with a bunsen burner. We have an inductive argument of the form:

N samples of sodium have been observed to burn with an orange flame when heated with a bunsen burner.

All samples of sodium will burn with an orange flame when heated with a bunsen burner.

As it stands this is invalid, but it becomes a valid deductive argument if we add the following premise: whenever N As are observed to also be Bs then all As are Bs; and let A be 'sample of sodium' and B be 'things that burn with an orange flame when heated in a bunsen burner'.

This principle is general and will also allow us to infer that all bread is nourishing by observing that N samples of bread have been observed to be nourishing so far. Of course, as we learned in Chapter 1, we need to add to the principles that the observations of As must be made under a wide variety of conditions, and that no instance has been found to contradict the universal generalisation that all As are Bs. If we do this, then we will be able to infer such generalisations validly as follows:

N As have been observed under a wide variety of conditions and all were found to be Bs.
No As have been observed to be non-Bs.
If N observations of As under a wide variety of conditions have been made, and all were found to be Bs, and no As have been found to be non-Bs, then all As are Bs.

All As are Bs

This is valid because it is not possible for the premises all to be true and the conclusion false; however, the obvious problem with this is that we have not yet specified how big the number N needs to be. Whatever number we come up with is going to seem arbitrary and, moreover, our inductive reasoning will have the following extremely counterintuitive feature; we will have no reason to believe all As are Bs at all, no matter how much evidence we have until we reach the number of observations N and then suddenly we will have complete certainty that all As are Bs and further observations will be completely irrelevant. But why should any particular number of observations allow one to be certain? This problem can be avoided by weakening the conclusion so that it states that 'probably all As are Bs', and stipulating that the probability here is proportional to the size of N. (We shall return to this approach below.)

The other obvious problem is that we seem to lack any justification for the principle of induction that is proposed. It does not seem to be an analytic truth (a relation among our ideas) because its negation is not a contradiction, but rather a synthetic proposition (a matter of fact). So if Hume is right it must be justified by experience and then we are back to the circularity problem again.

However, perhaps Hume is wrong and some synthetic truths can be known *a priori*. This is the response to the problem of induction inspired by Kant's idea that certain principles can be known to be true *a priori* because they are, in fact, descriptive of the way our minds work and express preconditions for us to have any experience of the world at all. Kant argued that the principle that all events have causes, and perhaps also the specific laws to be found in Newton's physics are known in this way. In the eighteenth century, when Kant was writing, this may have seemed plausible because at the time Newton's laws were being applied to all kinds of celestial and terrestrial phenomena and were successful time and time again. The image of a clockwork universe in which every event follows from previous events with necessity and predictability according to the basic laws of mechanics was a great source of inspiration to scientists and philosophers, and indeed in the nineteenth century most philosophers were not too worried by the problem of induction. However, once Newtonian mechanics was found to be false because of the inaccurate predictions it gave for observations of bodies moving with very high relative velocities, and for the behaviour of very small and very large objects, the problem of induction acquired a new urgency. From the modern perspective, Kant's belief in synthetic *a priori* knowledge seems hopelessly optimistic.

(5) *Hume's argument is too general. Since it does not appeal to anything specific about our inductive practices, it can only be premised on the fact that induction is not deduction*

The point of this response is to argue that Hume's argument is supposed to apply to all forms of inductive inference but that the description Hume gave of our inductive practices was over simplistic. Hume claimed that in forming expectations about the future behaviour of things we have previously observed, we assume that the future will resemble the past. However, it is ridiculous to suggest that this is all there is to our inductive practices. Sometimes we need only observe something a few times before we conclude that it will always behave in a similar way; for example, when trying a new recipe one would conclude after two or three successful trials that the dish will usually be tasty in future, whilst on other occasions we are very cautious about inferring the future behaviour of things even after many observations. Furthermore, we may observe that certain events are repeatedly conjoined in past experience but not conclude that they will be in future; for example, I observe that all my breaths to date have been followed by further breaths but I do not infer that all my breaths will be followed by further breaths, because I fit this pattern into the rest of my inductive knowledge that includes the claim that all human beings eventually die. Hence, our inductive reasoning is more complex than Hume suggests and usually when we infer a causal connection it is because we have tested a regularity in various circumstances and found a certain stability to the behaviour of things.

Human beings and other animals are, in fact, much better at induction than they would be if they just used enumerative induction, and it is easy to see why: an animal that could only learn that something was dangerous by testing this many, many times would not survive for long; hence a child learns not to put his or her hand on a hot stove after a couple of unpleasant sensations and does not wait until it has repeated the observation over and over again. Indeed, even in science, sometimes a single experiment or a few observations is taken to provide sufficient evidence for a theory, as in the case of the famous experiment which confirmed the prediction of general relativity that the path of light would be bent by passing close to the Sun. Only a lunatic would suggest that we need to do some more experiments to confirm that the catastrophic effects of the nuclear bombs dropped on Hiroshima and Nagasaki would recur if someone tried the same thing in the future.

So it seems that if there really are such things as inductive inferences then they are more complicated than the enumerative induction that Hume considers. Of course, this shows only that we need to describe our inductive practices in more detail before considering whether or not they are justified, but nonetheless it is argued that Hume's argument does not give us any reason to doubt them just because they are inductive. This is a promising strategy that is currently popular among some philosophers but I suspect that Hume would argue that, however sophisticated and complex our inductive practices are, they will ultimately depend on the assumption that the future will resemble the past, and that hence, if that principle cannot be justified, our inductive practices cannot be justified.

(6) Induction is really (a species of) inference to the best explanation, which is justified

Inference to the best explanation, which is sometimes called *abduction*, is the mode of reasoning that we employ when we infer something on the grounds that it is the best explanation of the facts we already know. For example, when somebody doesn't answer the door or the phone, we usually infer that they are not at home because that best explains the data we have. Similarly, it is argued, in science hypotheses are often adopted because of their explanatory power, for example, the hypothesis that the continents are not fixed on the surface of the Earth but are very slowly drifting in relation to one another is adopted by geologists because it explains the common characteristics of some rocks that are now thousands of miles apart, and also some correlations between the shapes of different continents.

This is a very popular way of solving Hume's problem and the appeal to inference to the best explanation is very important in the context of the debate about scientific realism. In order to evaluate this strategy we will need to consider the nature of explanation and that will be one of the main tasks of Chapter 7. For now, note that this strategy is often combined with the next one, for it is argued that the positing of causal relations or laws of nature is justified because it is the best way of explaining the existence of stable regularities in how things behave.

(7) There really are necessary connections that we can discover

If there really are necessary connections between events then they will ensure that the regularities we observe will continue to hold in the future (because a necessary connection is one that could not be otherwise). This idea can be developed either in terms of laws of nature or in terms of causal powers. Hume assumes that we cannot observe the necessary connections that are supposed to constitute causal relations, and argues that, therefore, we cannot know about them at all, and hence that the inductive reasoning, which depends upon the postulation of them for its justification, is without any foundation. Similarly, a Humean view of laws says that there is nothing to a law of nature over and above some regularity in events. However, we might argue that we can know about necessary connections after all. One way to defend this would be to argue that necessary connections do not need to be directly observed despite what Hume says. As mentioned above, we might argue that we know about necessary connections by inference to the best explanation. Usually when we posit some causal connection or law of nature it is not just because we have observed some regularity in phenomena, such as objects falling when we drop them, but we have also some understanding of how stable the regularity is if we vary various conditions, for example, we drop things in air, in water, we add wings to them and we observe that smoke does not fall when dropped and so on. Again we will have to postpone a proper discussion of this strategy until later.

(8) *Induction can be inductively justified after all, because even deduction can only be given a circular (in other words, deductive) justification*

This is a more sophisticated version of the circular defence of induction that Hume considers and rejects. A common way of putting Hume's argument is as follows. Induction must be justified by either a deductive or an inductive argument. A deductive argument with the conclusion that induction is justified would only be valid if at least one of the premises assumes that induction is justified (as in strategy 4 above). On the other hand, an inductive argument will only persuade us that induction is justified if we already accept that inductive arguments support their conclusions. Hence, there cannot be a non-circular or non-question-begging defence of induction.

However, as was famously illustrated in a story by Lewis Carroll (1895), deductive inference is only defensible by appeal to deductive inference and yet that doesn't lead us to reject it as irrational, so why is induction any worse off? To see this, consider the following pattern of deductive inference; someone believes some proposition, p, and they also believe that if p is true then another proposition q follows, and so they infer q. What could they say to someone who refused to accept this form of inference? They might argue as follows; look, you believe p, and you believe if p then q, so you must believe q, because if p is true and if p then q is true then q must be true as well. They reply, 'OK, I believe p, and I believe if p then q, and I even believe that if p is true and if p then q is true then q must be true as well; however, I don't believe q'. What can we say now? We can only point out that if you believe p, and you believe if p then q and you believe if p is true and if p then q is true then q must be true as well, then you ought to believe q, but once again we are just forming an if . . . then . . . statement and insisting upon the mode of inference which, by hypothesis, the person we are seeking to persuade rejects. The upshot is that this fundamental form of deductive inference, which is called *modus ponens*, cannot be justified to someone who does not already reason deductively.

The suggestion is that it is impossible to give a non-question-begging defence of any form of inference. Perhaps, then, our strategy with the inductive sceptic ought to take account of this. Hence, we can offer an inductive defence of induction to reassure those who already employ induction that it is self-supporting, but we will give up on trying to persuade someone who completely rejects inductive inference that it is legitimate, on the grounds that such a task cannot even be carried out for deduction.

(9) Retreat to probable knowledge

This strategy amounts to modifying the principle of induction so that it only sanctions the conclusion that all As *probably* possess property B. All scientific knowledge, it is sometimes said, is merely probable and never completely certain; the more evidence we accumulate the more certain we become but there is no end point to this process and any hypothesis, no matter how well-supported, may be false after all. Although this response to the problem of induction begins by conceding that we can never be 100 per cent certain that a generalisation will continue to hold in the future, the probabilist argues that we can come very close to certainty and that is all we need for the justification of scientific knowledge. Some versions of this response involve a theory of *degrees of belief*, according to which belief is not an all or nothing matter but a matter of degree. Degrees of belief are usually associated with dispositions to bet at different odds; for example, if you have a degree of belief of 0.5 then you are likely to bet in favour of the hypothesis only when the odds offered for it being true rise above evens. (In the form of the theory of confirmation known as *Bayesianism*, this response has been given a precise mathematical form.)

However, note that Hume's conclusion is not merely that we cannot be certain of the conclusion of an inductive argument, but the much more radical claim that we can have no reason at all to believe it to be true rather than false. This is because we have no reason to believe in the uniformity of nature. The retreat to probable knowledge does not give us any new grounds to believe in the latter, so it does not seem to solve Hume's problem. Furthermore, usually judgements about probabilities are based on the observation of frequencies; for example, we might observe that two-thirds of the population of England have brown eyes and infer that the probability of someone in England whose eyes we have not yet seen being brown is approximately 66 per cent. However, the problem with inductive inferences, in general, is that we have no idea what proportion of the total number of instances we have observed. Indeed, universal generalisations entail an infinite number of observations and so any proportion that we observe, no matter how large, will always be a negligible fraction of the total. This is enough to show that the mere retreat to probabilism is insufficient to solve Hume's problem.

(10) Agree that induction is unjustified and offer an account of knowledge, in particular scientific knowledge, which dispenses with the need for inductive inference

This is the radical response to the problem of induction proposed by Karl Popper (1902–1994). We shall consider his views in the next chapter.

It should be noted that various combinations of strategies 1, 5, 6, 7, 8 and 9 are the most popular in contemporary philosophy. Hence, someone might argue that Hume's argument shows us not that induction is irrational but that something is wrong with his reasoning (the sophisticated version of strategy 1), that what is wrong is that his account of our inductive practices is too crude (strategy 5), that our inductive practices really depend on inference to the best explanation where the explanations in question involve the existence of causal relations or laws of nature (strategies 6 and 7), and that inference to the best explanation cannot be justified in a completely non-question-begging way, but then no form of inference can (strategy 8). To this we might add that we only ever end up with a high degree of belief rather than certainty and that this is the best we can achieve and is, moreover, psychologically realistic (strategy 9). Together, this amounts to a pretty strong response to the problem of induction, but even if we can solve or dissolve Hume's problem of induction we still need to provide some positive account of what it is for something to count as evidence in favour of a hypothesis. Such an account is called a *theory of confirmation* and there are several available (Bayesianism is probably currently the most popular among philosophers). The articulation of inductivism in the history of philosophy of science is closely tied to the development of increasingly

51

sophisticated mathematical theories of probability, and the increasing use of statistics in science. However, it is worth noting that, despite a long history, there is no generally agreed upon solution to the problem of induction. It is for this reason that the philosopher C.D. Broad (1887–1971) called induction the glory of science and the scandal of philosophy.

2.3   Inductivism and the history of science

The problem of induction is a significant difficulty for inductivism as a theory of scientific methodology; however, since the former also threatens most of our everyday knowledge we ought not to reject inductivism too hastily on that basis. If we can somehow solve or dissolve the problem of induction and vindicate inductive reasoning, then in principle a large number of observations may be used to justify belief in some generalisation or scientific law. However, we still need to ask whether the account of scientific method that we developed in the previous chapter is a plausible reconstruction of the method employed in the actual history of science (recall question (1) at the beginning of this chapter). If it is not then we face a dilemma: either we conclude that the history of science is not as it should be and that scientific knowledge is therefore not justified after all; or we conclude that inductivism must be mistaken as an account of the scientific method because it fails to characterise the methods that have been used in the production of our best scientific knowledge.

Obviously, if there are just a few cases of marginal scientific theories where the method employed to develop them does not fit the inductivist model then the former horn of the dilemma may reasonably be grasped. After all, we do not expect the history of science to be always ideal and clearly there are cases where the verdict of the scientific community itself is that some scientists have not followed the scientific method. However, in such cases this also gives us good reason to reject their theories, as in the case of the manifestly racist and sexist anatomy advocated by some scientists in the nineteenth century that modern scientists regard as completely bogus. On the other hand, the more the practice of science fails to fit the inductivist account of the scientific method, especially if cases of the development of what are taken to be among the best and most successful theories fail to fit the account, the more plausible it becomes to take this as evidence that the inductivist account is flawed.

There is a certain kind of circularity here. On the one hand we want to know whether what we take to be scientific knowledge is really justified, and on the other, any account of the nature of the scientific method that entails that most scientific theories are not justified at all is liable to be rejected for that reason. This circularity arises because most philosophers of science have some kind of prior commitment, although perhaps minimal and restricted, to the rationality of science and the justification of scientific knowledge (for example, as mentioned in Chapter 1, antirealists may limit this knowledge to a description of the phenomena and not believe that scientific theories are true descriptions of the underlying causes of what we observe). Hence, most philosophers of science think that certain core scientific generalisations, such as sodium burns orange when heated, or all metals expand when heated, are as justified as any empirical knowledge could be. From this perspective, the philosophy of science aims to articulate the nature and source of the justification that our best theories enjoy, and hence an account of the scientific method and the source of justification in science will be inadequate if it fails to apply to the development of theories that are regarded as our best examples of scientific knowledge, such as Newton's mechanics, Maxwell's theory of electromagnetism and so on.

The point about these theories is that they are used every day by engineers in numerous practical applications and, even though we know that they are only accurate to a certain extent and that they give answers that are quite wrong in certain cases, it is inconceivable that we could come to regard either as bad science. However, it is important to note that this attitude is born out of many years' experience of using these theories. I am not here claiming that we should have a prior commitment to the rationality of the practice of any particular current science, nor to the accuracy of all scientific theories. It is only with the benefit of hindsight and the ability to look back on the development of mechanics and electrodynamics over several centuries that one can be sure that these theories, like the basic principles of optics and thermodynamics among others, embody some reliable and robust generalisations about how things usually behave. Again, what

I have said in this section is not intended to suggest that we ought to believe in the literal truth of what these theories say about the causes and explanations of those generalisations, nor should we think that the predictions issued by such theories are immune from future improvement.

Given this, it is clear that, as in other areas of philosophy, we need to reach what is known as a 'reflective equilibrium' between our pre-philosophical beliefs and the results of philosophical inquiry. Consider the following analogy; in ethics we inquire into questions about the nature of the good and the general principles that will guide us in trying to resolve controversial moral issues, such as abortion and euthanasia. However, ethicists would reject any ethical theory that implied that the recreational torturing of human beings was morally acceptable, no matter how plausible the arguments for it seemed. In ethics we demand that accounts of the good do not conflict with our most fundamental moral beliefs, although we will allow them to force us to revise some of our less central moral views. So it is with the philosophy of science; accounts of the scientific method that entail that those scientists who produced what we usually take to be the best among our scientific theories were proceeding in quite the wrong way will be rejected, but we will allow that an account of the scientific method can demand some revisions in scientific practice in certain areas. Indeed, it is permissible that we might conclude that most current science is being done very badly, or we might even conclude that most scientists are bad scientists; nonetheless, we ought not to conclude that our best science is bad science.

Hence, philosophy of science needs to be informed by careful work in the history of science and not just by accepting scientists' own pronouncements about how their work proceeds. In fact many histories of science – for example, of the discoveries of Galileo, Newton and the discovery of vaccination by Edward Jenner (1749–1823) – have been written from an inductivist perspective. Newton famously claimed not to make hypotheses, but to have inductively inferred his laws from the phenomena. It will be instructive briefly to consider the development of Newton's theory to see if it fits with the inductivist model.

In his celebrated *Principia* (the full title translates as *The Mathematical Principles of Natural Philosophy*), Newton presented his three laws of motion and his law of universal gravitation, and went on to use them to explain Kepler's laws of planetary motion, the behaviour of the tides, the paths of projectiles (such as a cannon ball) fired from the surface of the Earth, and many other phenomena. The law of gravitation stated that all massive bodies attract each other with a force (F) that is proportional to the product of their masses (m_1m_2) and is inversely proportional to the square of the distance (r) between them.

$$F = \frac{m_1m_2G}{r^2} \quad \text{(where } G \text{ is a constant)}$$

(This means that two bodies that are 10 m apart experience a force that is 100 times less than two equally massive bodies that are 1 m apart.) Newton makes a distinction between the law itself and some account of the cause or explanation of the law, and claims that his law has been inferred from the data, but also that, because no such inference leads to an account of what causes the gravitational force in accordance with the law, he suspends judgement as to what the cause might be. Indeed, Newton says that 'hypotheses', by which he means statements that have not been inferred from observations, have no place in 'experimental philosophy', being merely speculative.

A major problem with Newton's account of his own discoveries was famously pointed out by the historian and philosopher of science Pierre Duhem (1861–1916), namely that Kepler's laws say that the planets move in perfect ellipses around the Sun, but because each planet exerts a gravitational force on all the others and the Sun itself, Newton's own law of gravitation predicts that the paths of the planets will never be perfect ellipses. So Newton can hardly have inferred his laws directly from Kepler's if the latter are actually inconsistent with the former. Now consider Newton's first law, which states that every body will, unless acted upon by an external force, maintain its state of uniform motion (if it is already moving) or will remain at rest (if it is not). We have never been able to observe a body that is not acted upon by some external force or other, so again this law cannot have been inferred directly from the observational data. Furthermore, Newton introduced new theoretical concepts in his work. In particular, the notions of mass and force are both made precise and quantitative in the *Principia* and feature in the law of gravitation. However, Kepler's laws relate positions, distances, areas, time intervals and velocities and make no mention of forces and masses. How could a law, which is stated in terms of these theoretical concepts, be inferred from data where they are entirely absent?

Another historical example that is often taken to support inductivism is Kepler's discovery of his laws of planetary motion. Between 1576 and 1597, the astronomer Tycho Brahe (1546–1601) made thousands of observations of the planets, and Kepler used this data to produce his three laws, so it seems that here at least we have a case where a theory was inferred from a mass of observational evidence. However, Kepler was unable just to read off his laws from the data, rather he was motivated to search for a reasonably simple pattern to planetary motion by his somewhat mystical (Pythagorean) belief in a mathematically elegant form to the motion of the planets, which he thought of as the harmony of the spheres. There are numerous other examples of creative thinking in science where scientists certainly did not derive their theories from the data.

2.4 Theory and observation

Consider the requirement that before making an inductive inference we must examine the phenomena in question in a wide variety of conditions. Now, there are many cases of scientific laws and generalisations that were thought to be true without exception, but were then later found to be false when tested in certain situations. Newtonian mechanics is a prime example, since it is completely inaccurate when applied to things moving at very high relative velocities, yet it had been tested at lower speeds millions of times and always found to be pretty accurate. How do we know in advance what circumstances are significantly similar and different? Of course, we assume that it doesn't make any difference if the experimental device we are using is painted red or green but how do we know it doesn't? Similarly, we do not expect it to make any difference to whether a metal expands on heating whether we test this on one day or a year later, or we do it in the northern hemisphere or the southern hemisphere.

Obviously we rely upon background knowledge in deciding which circumstances to vary and which not to vary. If we are testing to see if all metals expand when heated, we think that it may be relevant whether we use a different type of metal, how we heat the metal, and how pure the sample is, but not whether the experimenter's name has an 'e' in it or in what order we test the metals. Here, we are using our background knowledge of what factors are causally relevant. What the experimenter's name is makes no difference to experiments we have carried out in the past so we do not expect it to make any difference to the next experiment. The accuracy of experimental techniques depends upon being able to detect and screen out extraneous influences. If we are doing basic mechanics with billiard balls we try and use a very smooth and flat surface to minimise the effect of friction. We may go on to study such systems in a vacuum to minimise air resistance. This process is called 'idealisation'. Often science proceeds by studying ideal systems where various complicating factors are not present, and then applying the derived laws to real systems and modifying them as appropriate.

Bacon recommended that we free our minds of all preconceptions when undertaking scientific inquiry, but is this possible and is it even desirable? We have seen how, to be plausible as an account of the scientific method, inductivism must admit that we need to use background knowledge to screen out causal factors in which we aren't interested. It may have seemed okay to start from scratch in Bacon's time in order to avoid being misled by the received Aristotelian wisdom that had become dogmatic and unproductive, but nineteenth and twentieth century scientists were building upon well-established and complex theories. They wanted to consolidate and extend that success and not ignore it when investigating new domains. So they needed to use the theories of optics to help build telescopes to study stars and microscopes to study cells. Modern science is so complex and developed it is absurd to suggest that a practising scientist has no preconceptions when undertaking research. Scientists need specialised knowledge to calibrate instruments and design experiments. We cannot just begin with the data, we need guidance as to what data are relevant and what to observe, as well as what known causal factors to take into account and what can safely be ignored.

There is another problem with inductive inference that we face even if we could show that the future is like the past. The problem was discovered by Nelson Goodman (1906–1998) and is known as the 'new riddle of induction'. Goodman argued as follows: suppose that the future will be like the past; we observe that every emerald we have ever come across has been green and so we infer that all emeralds are green. This is an exemplary case of enumerative induction where the generalisation is supposed to be supported or justified by the observation of a large number of instances consistent with it and none that contradict it, and suppose too that we have observed emeralds under a wide variety of conditions. Now consider the property 'grue', where a thing is grue just in case it is observed prior to the year 2001 and is green, or it is only observed after 2001 and is blue. Now all the emeralds we have observed up to now have been grue by this definition, and hence all the data we have supports the generalisation 'all emeralds are grue' just as much as it supports the generalisation 'all emeralds are green'.

Of course, the predicate 'grue' is artificial but Goodman's point is that we need some way of distinguishing those predicates with which we can legitimately make inductive inferences (call these 'projectible predicates'), from those predicates which we cannot legitimately make inductive inferences with (call these 'non-projectible predicates'). Goodman's problem remains even if we solve the ordinary problem of induction, and it also shows us that we need to say more about observation. On the simple model of observation we have assumed, it is just a matter of setting up some experiment and recording what happens objectively. But the possibility of grue type predicates means that we will get into trouble if we record our observations in the wrong language. (We shall return to the problematic relationship between theory and observation later.)

2.5   Conclusions

The general lessons to be learned from the history and practice of science are as follows:

(1) Sometimes new theories refine our understanding of the data we already have and so, in general, the former cannot be simply read off or inferred from the latter. For example, we come to regard the deviations of the paths of the planets from perfect ellipses not

as observational errors but as revealing the effects of the planets' gravitational attractions between themselves.

(2) The history of science has often involved the introduction of new concepts and properties that could not have been simply inferred from the data.

(3) Theories guide us in deciding what to observe under what conditions and especially in the case of modern science; presuppositionless observation would be detrimental even if it were possible. The relationship between theory and observation is much more complex than it seems at first sight.

(4) Many different influences (dreams, religious beliefs, metaphysical beliefs, and so on) may inspire a scientist to propose a particular hypothesis other than the data he or she already knows about.

So it seems that the model of science presented at the end of the previous chapter, which the reader may have taken to be quite natural and may even have been explicitly taught at school, is inadequate. In the next chapter we will consider the influential rival account of the nature of science and the scientific method advocated by Popper.

—·◦⊜◦·—

Alice: I can't give you a reason to follow the principle of induction, but that doesn't matter because it is impossible to get someone to follow any form of argument if they just refuse to. The fact remains that the vast majority of people think it is perfectly reasonable to base expectations of the future on past experience.

Thomas: That's it? So basically you're saying that most people use induction and those that don't are mad and you can't reason with them. What makes you think you're the sane one?

Alice: The thing is, it doesn't really matter either way. Sometimes there is no way of persuading someone who refuses to believe something that everyone else knows is justified. For example when someone is in denial about something. You know people who can't admit that they are an alcoholic, or that the person they are seeing is cheating on them, when everyone else thinks it is obvious. The stupid thing is the sceptic about induction gets proved wrong all the time, every time they step and gravity pulls them down to Earth.

Thomas: But lots of the time you can't predict how the future will be, and the patterns of the past are broken.

Alice: All I'm saying is that the fact that induction can't be justified to someone who doesn't reason that way doesn't mean that those of us who do can't know that it is generally reliable and justifies our scientific knowledge. Take those people who joined that religious cult that thought the world would end in 1999, and all killed themselves at some appointed hour to join a spaceship near that comet.

Thomas: That was beautiful that comet.

Alice: It was, and we don't need to think it's anything other than a natural phenomenon to appreciate that, just like we don't need to think that a rainbow is something other than the scattering of light waves by their passage through air that has lots of small droplets of water suspended in it. A comet is a bundle of frozen ammonia and water with a few other elements thrown in, in orbit around the sun like the rest of us. It is basically just a rock reflecting light not a chariot of some god or an alien spacecraft. We know this because we have theories that have been confirmed by predicting such phenomena in the past.

Thomas: So you say, but you can't just read off the right theory from what you see.

Alice: Well you can argue all you like but I am going to carry on believing the scientists and not the people who tell me that the world will end and that I had better repent, and give them all my money. By induction, I know that they are very probably wrong, and the fact that I can't convince them doesn't mean they aren't all off their heads.

Thomas: I take your point, but look, what I said in the first place was that there is no more to the scientific method than trial and error. I try and learn by my mistakes, so if you want to call that induction then I agree that I use it but that doesn't get us any closer to atoms and all that. You still haven't explained to me how you get from the fact that we all have to use induction sometimes, to believing all that stuff about the Big Bang. Anyway, I think the point about cultists and people like that is that they aren't prepared to abandon their beliefs in the face of the evidence. They just make up some just-so story to explain why they got it wrong and carry on regardless. The only thing that is good about science is an attitude of scepticism towards the traditional dogma.

—·◦◯◦·—

Further reading

Hume

Hume, D. (1963) *An Enquiry Concerning Human Understanding*, Oxford: Oxford University Press.
Woolhouse, R.S. (1988) *The Empiricists*, chapter 8, Oxford: Oxford University Press.

Induction

Ayer, A.J. (1956) *The Problem of Knowledge*, chapter 2, Harmondsworth, Middlesex: Penguin.
Goodman, N. (1973) *Fact, Fiction and Forecast*, Indianapolis: Bobbs-Merrill.
Papineau, D. (1993) *Philosophical Naturalism*, chapter 5, Oxford: Blackwell.
Russell, B. (1912) *The Problems of Philosophy*, chapter 6, Oxford: Oxford University Press.
Swinburne, R. (ed.) (1974) *Justification of Induction*, Oxford: Oxford University Press.

Inductivism and the history of science

Achinstein, P. (1991) *Particles and Waves*, Oxford: Oxford University Press.

3

———·❍·———

Falsificationism

One reason for wanting a theory of scientific method is so that we can ascertain whether scientific knowledge is justified and, if so, what its limits are. This may be important for interpreting scientific results about whether there is a risk associated with eating certain foods or releasing genetically engineered organisms into the environment. It may also be important for evaluating whether scientists' theories about the origin of the universe or the nature of matter are true or merely good guides to what we observe. Even if scientific theories, such as Newtonian mechanics, are recognised by all sides to be extremely reliable for predicting all kinds of phenomena, it remains an open question whether our best scientific theories also accurately describe unobservable entities that cause what we observe.

However, there is another reason for seeking an account of the scientific method, namely that if we have such an account we may be able to use it to decide whether some theory or discipline is scientific or not. In the United States of America, for example, there is a law that bans the state from establishing any particular religion. This law has been interpreted as prohibiting the teaching in state schools of the biblical account of God's creation of the world, animals and human beings. Yet, some of those who adhere to this doctrine call it 'creation science'. They argue that since their interpretation of the biblical account of the creation is a scientific theory it ought to be taught to students as part of the science curriculum. Many people disagree with the claim that the doctrine of the creation is genuinely scientific, although of course they admit it is possible to adopt the style and superficial appearance of science. It is therefore of considerable legal, political and religious importance whether it really is science, and this means that some account of the nature of science is needed. No matter how much self-styled creation scientists cite their alleged empirical evidence for the Garden of Eden, Noah's ark and the flood and other events of the Bible, most geologists and biologists are convinced that all the scientific evidence points to the Earth and the life it supports having been in existence for millions rather than thousands of years. (Whether or not God created the universe is another matter.) But even if they are right, is creation science just bad science rather than non-science?

Other allegations that particular theories or practices are pseudo-scientific are very much a part of contemporary scientific and political debate. For example, some scientists and philosophers have alleged that the notion of an 'intelligence quotient' (IQ) and the testing of it is pseudo-scientific (which means 'is claimed to be scientific but is not'), yet this and other forms of psychometric testing are used by schools, employers and medical agencies. Sometimes within a particular scientific discipline dissidents are labelled as pseudo-scientific. For example, an issue of the popular science magazine *New Scientist*, which I happened to read while I was writing this chapter, had an article on why some researchers were thinking of boycotting the 2000 World Aids Conference in South Africa. Some scientists think that the government there is neglecting its responsibilities by not funding the use of certain AIDS drugs and by questioning the widely held belief that AIDS is caused by the HIV virus. Professor Malegapuru Makgoba of the Medical Research Council of South Africa is quoted as saying that South Africa is becoming 'fertile ground for pseudo-science' (*New Scientist*, 29 April 2000: 15). By the way, there was also an issue around the same time on creation science which stated unequivocally 'science and religion inhabit different domains' (*New Scientist*, 22 April 2000: 3), yet if the Bible makes statements about the creation of the Earth how can this be? There seems no avoiding the fact that sometimes religious doctrines may conflict with accepted scientific theories, so if the former are dressed up in the guise of scientific theories, they need to be evaluated as such. But how can we tell whether creation science is genuine science or not? For each of parapsychology, acupuncture, astrology, homeopathy and many other practices, there have been people who have claimed the practice is scientific and others who claim it is not. Should publicly funded health, education and legal institutions be teaching and using such practices? Given the status science and scientists enjoy in contemporary life, it ought to be clear that deciding whether something is scientific or not will often be a decision with significant consequences for people's lives.

In the previous chapter, I argued that the simple account of the scientific method that was presented in Chapter 1 was inadequate. The problem of induction certainly shows that the justification of scientific knowledge is problematic and that there is a need for a precise theory of confirmation if any form of inductivism is to be defensible. However, the problem of induction also casts doubt on pretty much all of our empirical knowledge, even of everyday and trivial facts, such as that bread is nourishing or that salt placed in water will dissolve. Hence, someone wedded to naïve inductivism may be inclined to say that the problem of induction cannot be sufficient to refute it since we will be obliged to abandon so much common sense with it. Nonetheless, naïve inductivism also seems to be factually incorrect as an account of how many scientific theories were actually developed. Furthermore, the idea of presuppositionless observation seems both impossible and undesirable. It seems that naïve inductivism cannot deliver the demarcation of science from non-science because it does not give us a plausible account of how science develops, and it forces us to reject our core intuition that a theory such as Newtonian mechanics is an example of a good scientific theory. In this chapter, we will consider an alternative theory of the nature of the scientific method, and the grounds for the demarcation of science from non-science, called **falsificationism**. The discussion of falsificationism will suggest ways in which we can improve on naïve inductivism while retaining some of the core intuitions behind it, and at the end of the chapter I will formulate a more sophisticated inductivism.

3.1 Popper and the critique of Marxism and psychoanalysis

Karl Popper had a considerable influence on philosophy of science during the twentieth century and many scientists took up his ideas.

As a result, he was made a member of the Royal Society of London, which is one of the most prestigious scientific associations. In fact, Popper's falsificationism is probably now more popular among scientists than it is among philosophers. Popper also played an important role in the intellectual critique of Marxism, and his books *The Poverty of Historicism* and *The Open Society and Its Enemies* are still widely read by political theorists today. His interest in philosophy of science began with the search for a demarcation between science and pseudo-science. He tried to work out what the difference was between theories he greatly admired in physics, and theories he thought were unscientific in psychology and sociology, and soon came to the conclusion that part of the reason why people erroneously thought that mere pseudo-sciences were scientific was that they had a mistaken view about what made physics scientific.

The main battleground of the debate about demarcation is social science. The ideal of social science was a product of the eighteenth century, which was a time of general intellectual excitement and enthusiasm for the success of Newtonian physics and the other new sciences of chemistry, physiology and so on, that had recently advanced and expanded rapidly. Various thinkers suggested that the logical next step was the application of the same methods to the discovery of the laws that governed human behaviour and the way societies functioned. This period in intellectual history is known as 'the age of enlightenment and reason' and it was characterised by a profound optimism about what could be achieved if only human beings could learn to organise themselves on a rational basis in accordance with a genuine science of society. At the time, when Popper formed his views about science, in the early part of the twentieth century, there were theories of the social and psychological nature of human beings that were claimed by their adherents to fulfil the Enlightenment promise of a genuine science of society and human behaviour. Marxism and psychoanalysis were prominent among these theories.

At the funeral of Karl Marx (1818–1883), his friend and collaborator Frederick Engels (1820–1895) said that just as Darwin had discovered the scientific principles underlying the development of species, so Marx had discovered the scientific principles underlying the development of societies. Similarly, Sigmund Freud (1856–1939) claimed that his own discoveries were comparable to those of Copernicus and Darwin, and considered his theories of sexual repression, and of ego, id and superego and so on to be fully scientific. For various reasons, Marxism and psychoanalysis are both widely perceived (perhaps correctly) as somewhat discredited today; however, many of the twentieth century's greatest intellects were influenced by one or other of them, and it is arguable that their effect on the history of the twentieth century was profound. When he was young, Popper was attracted by both Marxism and psychoanalysis yet fairly quickly he grew disillusioned with them. He came to regard both as pseudo-scientific and set about trying to explain what it was about them and the way they were practised that led him to this view.

Popper realised that it was easy to think of both these theories as very successful sciences if one assumed that scientific knowledge proceeds, and is justified, by the accumulation of positive instances of theories and laws. On this view, as we have seen, the justification of a law such as all metals expand on heating would be a matter of there being many cases of particular metals that expanded when heated. Marxists and psychoanalysts both had numerous examples of phenomena that were instances of their general principles. The problem, as far as Popper was concerned, is that it is just too easy to accumulate positive instances which support some theory, especially when the theory is so general in its claims that its seems not to rule anything out. Popper certainly seems to be on to something here. People are often disdainful of horoscopes precisely because they are so general it is hard to see what would not count as supporting evidence for their claims. For example, your horoscope might read 'you will have money worries shortly'. There are not many people who don't regularly have money troubles. Similarly, suppose your astrological chart says that you lack confidence, or that you are friendly but sometimes shy. Very few people can claim to be confident in all respects or never to feel shy in some circumstances. Of course, I am not arguing here that astrology is a psuedo-science, and I am sure that some astrologers say things that are much more specific. The point is that if someone does make such vague pronouncements, it is surely not enough to make their theory scientific that many instances can be found that conform to it. Hence, Popper thought that theories that seem to have great explanatory power are suspect precisely because so much can be explained by them.

Similarly, Popper says that many adherents of Marxism and psychoanalysis are over-impressed with explanatory power and see confirmations everywhere. He argues that Marxists see every strike as further evidence for the theory of class struggle, and that psycho-analysts treat every instance of neurosis as further evidence for Freud's theories. The trouble with their theories is they do not make precise predictions, and any phenomena that occur can be accounted for. Indeed, both theories are able to explain evidence that seems at first sight to refute them. So, for example, various measures to safe-guard the safety and welfare of workers were introduced in England in the nineteenth century and this fact would seem to conflict with Marxism, according to which the ruling class has no interest in ensuring decent living and working conditions for the poor. Yet some Marxists have argued that, in fact, the introduction of the poor laws and so on confirm Marxism because they show that the capitalists were aware of the imminence of the proletarian revolution and were trying to placate the workers in order to stop or delay it.

In the case of psychoanalysis, Popper gives two different examples of human behaviour; the first is that of a man pushing a child into water intending to drown it and the second is of a man jumping in and sacrificing his life to save the child. Freud could explain the first by positing that the man suffered from repression, and the second by saying he had achieved sublimation. Alfred Adler (1870–1937) could explain the first by saying that the man suffered from feelings of inferiority and so needed to prove to himself that he could commit the crime, and the second by saying that the man also suffered from feelings of inferiority but needed to prove to himself that he was brave enough to rescue the child. Popper's complaint then is that the central principles of these theories are so general as to be compatible with any particular observations and too many of those who believe them cannot even imagine circumstances under which they would be empirically refuted because they are like a lens through which they view the world.

So, in general, Popper's worry about the idea that confirmation is fundamental to the scientific method is that if you are in the grip of a theory it is easy to find confirming instances, especially if the theory is one that is vague and general. By contrast, Popper was particularly impressed by the experimental confirmation of Einstein's general theory of relativity in 1917. The latter predicted that light passing close to the Sun ought to have its path bent by the Sun's gravitational field. The admirable thing about the theory as far as Popper was concerned was that it made a prediction that was very risky, which is to say that could easily have turned out to be false. There are plenty of other examples of such potentially falsifying, and therefore risky, predictions made by scientific theories. For example, Newton's theory predicted the return of Halley's comet during 1758, and made numerous other precise predictions for the behaviour of mechanical systems. However, the most compelling types of prediction for Popper were so-called novel predictions, which were predictions of new types of phenomena or entities. The example from general relativity mentioned above is of this kind. Another famous example is Dmitry Mendeléeff's (1834–1907) prediction of the existence of the previously unknown elements of gallium and selenium derived from the structure of the Periodic Table of the elements. Popper thought that the issuing of novel and risky predictions was a common characteristic of scientific theories and that this, combined with scientists' willingness to reject a theory if its predictions were not observed, was what made science so intellectually respectable.

So Popper argued that the 'confirmation' that a theory is supposed to get from observation of an instance that fits the theory, only really counts for anything when it is an instance that was a risky prediction by the theory; that is, if it is a potential falsifier of the theory. He thought that the impressive thing about genuine scientific theories is that they make precise predictions of surprising phenomena and genuine scientists are prepared to reject them if their predictions are not borne out by experiments. Not only are Marxism and psychoanalysis too vague to be subject to refutation by experience, but furthermore, Marxists and psychoanalysts are also sometimes inclined to side-step intellectual critique because their theories explain why people will oppose them. If one rejects Marxism one may well be accused of having a class interest in maintaining the capitalist system; similarly, someone who strongly opposes psychoanalysis may well be accused of being repressed. Of course, it is possible either or both of these claims are correct in many or even all cases; the point is just that these theories seem to foreclose the possibility of criticism, and it was this characteristic that Popper considered anathema to science. Hence, Popper came to the view that it is not confirmation but falsification that is at the heart of the scientific method.

3.2 Popper's *solution* to the problem of induction

Popper's solution to the problem of induction is simply to argue that it does not show that scientific knowledge is not justified, because science does not depend on induction at all. Popper pointed out that there is a logical asymmetry between confirmation and falsification of a universal generalisation. The problem of induction arises because no matter how many positive instances of a generalisation are observed it is still possible that the next instance will falsify it. However, if we take a generalisation such as all swans are white, then we need only observe one swan that is not white to falsify this hypothesis. Popper argued that science is fundamentally about falsifying rather than confirming theories, and so he thought that science could proceed without induction because the inference from a falsifying instance to the falsity of a theory is purely deductive. (Hence, his theory of scientific method is called *falsificationism*.)

Popper argued that a theory that was, in principle, unfalsifiable by experience was unscientific. Examples of statements that are not falsifiable include:

Either it is raining or it is not raining.
God has no cause.
All bachelors are unmarried.
It is logically possible that space is infinite.
Human beings have free will.

Clearly, no number of observations would be sufficient to refute any of these theories. Now as we have seen, Popper also thought that a theory like 'all neuroses are caused by childhood trauma' was unfalsifiable and so unscientific. On the other hand, he though that Marxism was falsifiable and so potentially scientific, since it predicted an internationalisation of the working class and a communist revolution. Popper just thought that Marxists were clinging on to a refuted theory and so were bad scientists. (It should be noted that here the distinction between being a bad scientist and a pseudo-scientist becomes somewhat unclear.) On the other hand, the examples of scientific theories we have considered are falsifiable because there are observations that are inconsistent with them. If we were to observe a metal that did not expand when heated we would know that the generalisation 'all metals expand when heated' was false. Similarly, if light did not obey the law of reflection we could observe this, and if bodies do not obey Newton's law of gravitation we ought to be able to observe their deviations from its predictions.

Having distinguished between falsifiable and unfalsifiable hypotheses, Popper argues that science proceeds not by testing a theory and accumulating positive inductive support for it, but by trying to falsify theories; the true way to test a theory is not to try and show that it is true but to try and show that it is false. Once a hypothesis has been developed, predictions must be deduced from it so that it can be subjected to experimental testing. If it is falsified then it is abandoned, but if it is not falsified this just means it ought to be subjected to ever more stringent tests and ingenious attempts to falsify it. So what we call confirmation is, according to Popper, really just unsuccessful falsification:

> [F]alsificationists like myself much prefer an attempt to solve an interesting problem by a bold conjecture, even (and especially) if it soon turns out to be false, to any recital of a sequence of irrelevant truisms. We prefer this because we believe that this is the way in which we can learn from our mistakes; and that in finding that our conjecture was false we shall have learnt much about the truth, and shall have got nearer to the truth.
>
> (Popper 1969: 231)

This is why Popper's methodology of science is often called the method of 'conjectures and refutations' (and indeed that was the name of one of his books). 'Bold' conjectures are those from which we can deduce the sort of novel predictions discussed above. According to Popper, science proceeds by something like natural selection and scientists learn only from their mistakes. There is no positive support for the fittest theories, rather they are just those that repeatedly survive attempts to falsify them and so are the ones that are retained by the scientific community. It is always possible that our best theories will be falsified tomorrow and so their status is that of conjectures that have not yet been refuted rather than that of confirmed theories. Popper thought that it is here that the intellectual corruption of Marxists and psychoanalysts lies for whether or not their theories are falsifiable – they do not state clearly the conditions under which they would give up their theories. It is this *commitment* to their theories that Popper thinks is unscientific. In fact, he demands of scientists that they specify in advance under what experimental conditions they would give up their most basic assumptions. For Popper, everything in science is provisional and subject to correction or replacement:

> [W]e must not look upon science as a 'body of knowledge', but rather as a system of hypotheses which in principle cannot be justified, but with which we work as long as they stand up to tests, and of which we are never justified in saying that we know they are 'true' or 'more or less certain' or even 'probable'.
>
> (Popper 1959: 317)

The view that knowledge must be certain, a matter of proof and not subject to error has a long history in philosophy. However, from Popper we learn that we should always have a critical attitude to our best scientific theories. The history of science teaches us that even theories that in their time were considered highly confirmed and which enjoyed a huge amount of empirical success, have been shown to be quite mistaken in certain domains. Overall, the history of science has seen profound changes in fundamental principles. For example, the Newtonian conception of a world of material particles exerting gravitational forces on each other and subject to the laws of Newtonian mechanics whizzing around in the void was replaced by the idea of a field that was present at all the points of space. Special relativity and quantum mechanics meant that the basic laws of mechanics had to be revised, and general relativity has led to radical changes in the way we view the universe and space and time. On a more mundane level, heat was once widely believed to be a material fluid ('caloric') that flowed unseen but felt, but now it is thought of as a manifestation of the kinetic energy of particles; whales are no longer regarded as fish, and the age of the Earth is now thought to be millions not thousands of years.

In the light of all this, it is not surprising that today not many people believe that any scientific theory is provable beyond all doubt. Popper fully endorses the philosophical position known as *falliblism* according to which all our knowledge of the world is provisional and subject to correction in the future. His theory of knowledge is thoroughly anti-authoritarian and this is linked to his critique of totalitarian systems of government. In his view, the programmes to create ideal societies proposed by the likes of Plato and Marx demanded rigid adherence to a single fixed ideology and the repression of all dissenting views. On the contrary, Popper thought that science flourished in an atmosphere where nothing is sacred and scientists can be extremely adventurous in the theories they propose. As his colleague Imre Lakatos (1922–1974) says, according to Popper, 'virtue lies not in caution in avoiding errors but in ruthlessness in eliminating them' (Lakatos 1968: 150). This accords with the familiar idea that scientists should be sceptical even about their own theories and should be ready to challenge any dogma if experiment demands it.

It is important to note that, unlike the logical positivists, Popper did not offer a way of distinguishing meaningful from meaningless statements and then argue that pseudo-science is meaningless. On the contrary, he thought that hypotheses that were not falsifiable could still be perfectly meaningful. Nor indeed did Popper argue that only what was falsifiable was helpful or productive even within science. Hence, he did not think that unfalsifiable metaphysical theories ought to be rejected altogether, for he recognised that sometimes scientists might be inspired to make interesting bold conjectures by beliefs that are themselves unscientific. So for example, many scientists have been influenced by their belief in God, or by their belief in the simplicity of the basic laws of physics, but clearly neither the proposition that God exists or that the fundamental structure of the world is simple is falsifiable by experience. Popper's theory of the scientific method allows such beliefs to play a role in scientific life even though they are not themselves scientific hypotheses.

Popper's main concern was to criticise pseudo-science because its adherents try to persuade people that their theories are scientific when they are not. It does not follow from the demarcation of science from pseudo-science that he proposed that there is anything wrong with a discipline or practice being non-scientific. In fact, Popper thought that both Marxism and psychoanalysis might embody important insights into the human condition; his point is just that they are not scientific, not that they are therefore not valuable. Obviously a strong case could be made for the value of religious beliefs, and it is perfectly possible for someone with religious faith and beliefs to accept a definite demarcation between science and religion (in fact I suspect this may be the case with many scientists).

As I pointed out above, the falsificationist does not view all scientific theories equally. Some theories are falsifiable but the phenomena they predict are not interesting or surprising. Hence, the hypothesis that it will be sunny tomorrow is certainly falsifiable though it is not of great value within science. Recall that the hypotheses that Popper prizes above all others are bold conjectures that make novel predictions. In fact, Popper believed that hypotheses can be compared to see which is more falsifiable: for example, take the hypothesis (1) that all metals expand on heating; it is more falsifiable than the hypothesis (2) that copper expands on heating, because the former hypothesis is inconsistent with more observation statements, in particular, it is inconsistent with observation statements about particular bits of iron and silver not expanding when heated as well as those that just concern copper. In this case, the set of all potential falsifiers of (2) is a subset of the set of all potential falsifiers of (1), and hence (1) is more falsifiable than (2).

Popper thought that theories could be ranked according to their degree of falsifiability and that this is the true measure of their empirical content. The more falsifiable a theory is the better it is because if it is highly falsifiable it must make precise predictions about a large range of phenomena. This seems to accord with an intuitive idea of what makes a particular scientific theory a good one. Scientists ought to aim to develop theories that are as falsifiable as possible which means the theories need to be both precise and have a broad content. For example, a hypothesis such as 'metals change shape when heated' is falsifiable and broad in scope but not precise enough to be highly falsifiable, while a hypothesis such as 'this piece of copper expands on heating' is pretty precise but of narrow scope.

Ideally, from the falsificationist point of view, science ought to consist of hypotheses that apply to a wide range of phenomena, but also make precise quantitative predictions. This is the situation with many of our best scientific theories, for example, Newton's mechanics gives precise predictions for a wide range of phenomena, from the motions of comets in the heavens to the paths of cannon balls near the surface of the Earth. Popper also argued that new theories ought to be more falsifiable than the theories they replace. This certainly fits with many episodes in the history of science; for example, Newton's theory was more precise than Kepler's which it succeeded, the theory of relativity improved upon the predictions of both Newtonian mechanics and Maxwell's electromagnetic theory, and so on. It seems that some of the basic ideas of falsificationism do accord with some of our intuitions about science.

3.3   The context of discovery and the context of justification

The attentive reader may have noticed a striking difference between naïve inductivism and falsificationism, namely that the former offers an account not just of how to test a scientific theory, but also an account of how scientists ought to generate them. So recall that Bacon's new inductive logic tells us how to begin our investigation of some range of phenomena, and the production of generalisations and laws is supposed to be an automatic outcome of the mechanical operation of the method. For a long time in the history of science it was widely believed that laws ought only to be admitted if they were actually derived from experimental data, and Newton himself claimed that he did not engage in speculation but simply deduced the laws of mechanics from the results of observations. However, as was explained at the end of the last chapter, we now know that in most of the interesting cases this is just not possible. Even Newton's laws cannot be simply read off the data, and claims of the sort he made are now not taken seriously. If there is one thing that has been learned from the twentieth century debates about scientific method it is that the generation of scientific theories is not, in general, a mechanical procedure, but a creative activity. If this is right, then when we are thinking about scientific methodology, perhaps we ought to make a distinction between the way theories are conceived and the subsequent process of testing them. In Popper's work, this distinction was absolutely pivotal because he thought that philosophy of science was really only concerned with the latter.

Popper was one of the first philosophers of science to emphasise that scientists may draw upon diverse sources of inspiration, such as metaphysical beliefs, dreams, religious teachings and so on, when they are trying to formulate a theory. He thought that none of these were illegitimate because he thought that the causal origins of a hypothesis were irrelevant to its status within science. The kind of speculation and imagination that scientists need to employ cannot be formalised or reduced to a set of rules. In a way this makes the sciences closer to the arts than they might otherwise seem. On the other hand, the sciences differ from the arts in being subject to testing by experience and this must be the final arbiter of any scientific dispute. Popper thought that the task of philosophy of science was to undertake the logical analysis of the testing of scientific theories by observation and experiment rather than to explain how theories are developed:

> [T]he act of conceiving or inventing a theory seems to me neither to call for logical analysis nor to be susceptible to it . . . the question of how it happens that a new idea occurs . . . may be of great interest to empirical psychology; but it is irrelevant to the logical analysis of scientific knowledge.
>
> (Popper 1934: 27)

In Popper's view then, there are two contexts in which we might investigate the history of science and the story of how certain theories come to be developed and accepted, namely the context of discovery and the context of justification. This view accords with an intuition about the autonomy of ideas from the people that have them. It is no argument against vegetarianism to point out that Hitler was a vegetarian; similarly it is no argument against Newtonian mechanics to point out that Newton was an alchemist, and had an obsessive interest in the apocryphal books of the Bible. On the other hand, it is no argument for pacifism to point out that Einstein was a pacifist. In general, the evidence in favour of a hypothesis is independent of who believes it and who doesn't, and whether an idea really is a good one is not at all dependent on whether it is a genius or a fool who first thinks of it. It seems plausible to argue that an evaluation of the evidence for a hypothesis ought to take no account of how, why and by whom the hypothesis was conceived. Some such distinction between the causal origins of scientific theories and their degree of confirmation is often thought to be important for the defence of the objectivity of scientific knowledge.

If we assume the distinction between the production of scientific theories and their subsequent testing, then we need not be troubled by the problems Bacon's theory of scientific method faced with the impossibility of freeing ourselves of all presuppositions when making observations, and the need for scientists to use background theories in the development of new ones. In fact, Bacon himself distinguished between 'blind' and 'designed' experiments and suggested that the latter were more useful in science because they will allow us to chose between two rival hypotheses that equally account for the data we have so far. The idea is that scientists faced with a choice between two seemingly equally good rival theories ought to construct an experimental situation about which the hypotheses will predict different outcomes. This is just the sort of thing Popper emphasised, and some people have argued that the standard accounts of Bacon's methodology of the sort I gave in Chapter 1 misrepresent his views and neglect the fact that Bacon anticipated what would later be called **hypothetico-deductivism**. This is the name given to the popular view that science is fundamentally about thinking up hypotheses and deducing consequences from them, which can then be used to test the theory by experiment. As I mentioned in Chapter 1, such experiments are often called 'crucial experiments', and a famous example is the experiment performed by French scientists in the eighteenth century to decide between Newton's theory of gravity and the theory preferred by those who followed René Descartes (1596–1650). The former predicted that the Earth would not be a perfect sphere but would be flattened at the poles by its own gravitational forces; the latter predicted the Earth would be elongated at the poles. The French sent expeditions to determine the dimensions of the Earth, and it was more or less as Newton's theory predicted. It is alleged that there are many such examples in the history of science, and hypothetico-deductivists believe that such experiments are of central importance for understanding scientific methodology. However, it has been argued that crucial experiments are, in fact, impossible; this will be the subject of the next section.

3.4   The Duhem problem

According to the account of falsificationism I have given so far, scientific theories are tested as follows: scientists deduce a prediction from a hypothesis and then if observation is not consistent with the prediction when the relevant experiment is performed the hypothesis is falsified. The way of thinking about falsification suggests the following schema to represent the relationship between a theory T and the observation statement that falsifies T:

$T \vdash e$ This says that T *entails* e, where e is something that can be decided by observation
$\neg e$ This says that e is false
$\neg T$ This says that T is false

For example, suppose T is the theory that all metals expand on heating, e is the statement that a particular sample of copper expands on heating. Clearly, T entails e, and so if e is false then T is false; the above argument is deductively valid.

However, in reality it is never possible to deduce any statement about what will be observed from a single hypothesis alone. Rather, hypotheses have to be conjoined with other assumptions about background conditions, the reliability of measurements, the initial conditions of a system and so on. This feature of the testing of scientific theories was recognised by Duhem who said: 'an experiment in physics can never condemn an isolated hypothesis but only a whole theoretical group' (Duhem 1906: 183). Consider the experimental test of Newtonian gravitational theory by the observation of the path of a comet. The law of gravitation alone will not predict any path for the comet. We need to assign values to variables representing the mass of the comet, the mass of the other bodies in the solar system and their relative positions and velocities, the initial position and velocity of the comet relative to the other bodies in the solar system, and the gravitational constant. We also need to employ Newton's other laws of motion. This will allow us to derive a prediction of the comet's future path that we can then test by observing its actual motion using a telescope. Suppose that the comet does not follow the path that Newtonian theory predicts; where do we locate the problem? It could be that the law of gravitation is false, or that one of Newton's other laws is false, or that we have one of the values for the mass of the other bodies in the solar system wrong, or that a mistake was made in observing the comet, or that the laws of optics which we think explain how the telescope works and why it is reliable might be wrong, and so on. Clearly, the falsification of a theory by an observation is not as straightforward as the above schema suggests.

Duhem discusses a real example that was widely considered to be a crucial experiment in optics. In the eighteenth century, there were two rival theories of the nature of light; one, due principally to Newton, according to which light consists of a stream of fast moving tiny particles, and the other, due principally to Christiaan Huygens (1629–1695), according to which light consists of a wavelike disturbance propagating through a unknown medium that permeates all space. Newton's theory predicted that the speed of light in water is greater than the speed of light in air. Eventually an experiment was devised such that light from the same source would pass through both water and air, and by the clever use of a rotating mirror the situation could be arranged so that the light would form two spots, one greenish the other colourless. If light travels faster in water than in air then the colourless spot ought to be to the right of the greenish one, and vice versa if light travels slower in water than in air. So we have a case where a statement describing something observable, namely 'the colourless spot appears to the right of the greenish one', can be deduced from a theory and we can try to falsify it. When the experiment was performed it was determined that the speed of light in water is in fact less than in air, and this was widely taken to refute Newton's theory, and to support the rival wave theory.

However, as Duhem points out, the situation is not so simple. Newton's theory, from which it follows that light travels faster in water than in air, includes a whole host of assumptions other than that light consists of particles. For example, Newton assumed that the particles of light attract and repel each other but that these forces are negligible unless the particles are very close together. It is all these hypotheses together that are inconsistent with the result of the experiment. So a more realistic schema for falsification would be as follows:

$(T\&A) \vdash e$ This says that T together with some set of auxiliary assumptions *entails e*

$\neg\, e$ This says that e is false

$\neg\, (T\&A)$ This says that the conjunction of T and the auxiliary assumptions is false

Now $\neg\, (P\&Q)$ is logically equivalent to $\neg P$ or $\neg Q$. (This should be obvious; if it is false that P and Q are both true, then either P is not true or Q is not true, or both.) So how do the scientists know whether T or one of the assumptions in the set A has been falsified by the experiment?

Duhem recognised that this problem was not widely appreciated. Whether or not people are thinking in falsificationist terms, people, perhaps even some scientists, often think that scientific hypotheses can be taken in isolation and tested by experiment, to be either retained or discarded on that basis. In fact, says Duhem:

> [p]hysical science is a system that must be taken as a whole; it is an organism in which one part cannot be made to function except when the parts that are most remote from it are called into play, some more so than others, but all to some degree.
>
> (Duhem 1906: 187–188)

Furthermore, why can't we take an instance of falsification to be a refutation of the laws of logic rather than as refuting our hypothesis? A philosopher who argues that ultimately we could chose to abandon logic, rather than reject a physical theory in the face of falsifying evidence, is the American philosopher W.v.O. Quine (1908–2000). Quine argued that it would be quite reasonable to reject a law of logic, or change the meaning of our terms, if it was more convenient than rejecting a particular theory. Quine therefore rejects the distinction between analytic and synthetic truths that Hume, Kant and the logical positivists believed to be fundamental to epistemology (see 2.1, 5.3.1, and 6.1.3). A trivial example of such a change in the meaning of a term is that of the change in meaning of 'atom' which once meant something indivisible and now refers to a particular type of collection of smaller particles. When physicists discovered that atoms were divisible, they redefined 'atom' rather than abandoning the term altogether.

Whether or not Quine is right in his more radical conclusions, it is clear that Popper must grant that there is no such thing as a completely conclusive refutation of a theory by experiment. In fact, Popper admits this and argues that as well as a set of observation statements that are potential falsifiers of the theory, there must also be a set of experimental procedures, techniques and so on, such that the relevant group of scientists agree on a way in which the truth or falsity of each observation statement can be established. Hence, falsification is only possible in science if there is intersubjective agreement among scientists about what is being tested on any given occasion. Popper argues that, in proper scientific inquiry, whenever a high-level theoretical hypothesis is in conflict with a basic observation statement, it is the high-level theory that should be abandoned. Although Popper concedes that falsification of a high-level theory by an observation statement is not a matter of the evidence *proving* the theory to be false, he does argue that it is conclusive as far as the practice of science is concerned; intersubjectively testable falsification is, he says, final. If a hypothesis has enjoyed some empirical success in the past but is subsequently falsified, it must be abandoned and a new hypothesis should be proposed. The latter should explain whatever success was enjoyed by its predecessor, but it should also have extra empirical content that its predecessor does not have. It is in this way that true science avoids the deplorable state of affairs that occurs when a pseudo-scientific theory is falsified and its adherents simply introduce a new version of the theory to which arbitrary assumptions have been added to save it from falsification.

Some people have argued that because falsification is never completely conclusive there is not really the asymmetry between falsification and confirmation that Popper thought there was. This is a mistake because it is still the case that if the scientific community accepts the truth of a statement reporting the observation of a negative instance of some theory, for example, that some particular metal does not expand when it is heated, it is logically inconsistent for the community to believe the generalisation as well. On the other, hand there is nothing inconsistent in accepting the truth of a positive instance of the same generalisation and at the same time believing the generalisation to be false.

3.5   Problems with falsificationism

There are several problems with Popper's account of falsificationism. Some of this are specific to the details of the theory Popper first elaborated and so may be avoided by a more careful formulation or by revising some of the details. However, some are quite general and challenge the fundamental idea that it is possible to give an account of the scientific method without endorsing any kind of inductive inference. Below, some of the main criticisms of falsificationism are briefly explained.

(1)   *Some legitimate parts of science seem not to be falsifiable*

These fall into three categories.

(a)   *Probabilistic statements*

Science often seems to issue statements about the probability of some occurrence. For example, modern physics tells us that the half life of uranium 235 is 710,000,000 years, which means that the probability of one atom of uranium decaying in 710,000,000 years is one-half or that it is highly probable that if one starts with 1 kg of uranium then in 710,000,000 years 500 g of it will have decayed. However, such statements cannot be falsified because an experiment may produce an improbable outcome and that is consistent with the original statement – improbable things are bound to happen sometimes. Any statement about the probability of a single event is not falsifiable, so, for example, the probability that a particular coin toss will land heads is 1/2, but we cannot falsify that hypothesis by tossing the coin because the fact that the probability is 1/2 is consistent with the coin landing heads or tails on that occasion. This problem does not arise for probabilities that are defined over large populations; hence, the statement that the probability that a particular coin will land heads

50 per cent of the time during a million tosses would be considered refuted if the coin landed tails 90 per cent of the time. I won't say any more about probabilistic statements and theories except to point out that probability is a bit of a philosophical minefield for anyone, and that Popper did develop a detailed theory of probability whose merits we cannot assess here.

(b) Existential statements

Although Popper is right that a universal generalisation can be falsified by just one negative instance, many statements in science are not of this form. For example, scientific theories assert the existence of things like black holes, atoms, viruses, DNA and so on. Statements that assert the existence of something cannot be falsified by one's failure to find them. Of course, if a theory asserts the existence of something that we repeatedly fail to find in various circumstances then one has inductive grounds for thinking it won't be found in the future; however, falsificationism is supposed to allow us to do without inductive grounds for beliefs completely. This raises the question of the relationship between falsificationism and scientific realism. Popper is clear that belief in unobservable entities has often been an important influence on the ideas of scientists and has helped them generate highly falsifiable theories, such as the atomic theory of the elements that are central within physical science. However, his views on induction imply that one can never have positive grounds for believing in theoretical entities no matter how empirically successful the theories that posit them are. This contradicts the idea many people have that we have good reasons to believe that the entities to which our best current scientific theories seem to refer do in fact exist. We shall return to this issue later.

(c) Unfalsifiable scientific principles

It is arguable that some unfalsifiable principles may nonetheless be rightly considered part of scientific knowledge. So, for example, the status of the principle of conservation of energy, which states that energy can take different forms but cannot be created or destroyed, is such that it is inconceivable to most scientists that an experiment could falsify it; rather, an apparent violation of the principle would be interpreted as revealing that something is wrong with the rest of science and it is likely that a new source, sink or form of energy would be posited. It has also been argued that the second law of thermodynamics, which states that the entropy of any closed system always increases, is of such generality that it is beyond falsification. Similarly, consider the principle that there is no 'action at a distance', in other words that all physical causation is mediated by local inter-actions. What this means is that whenever a distant event causes one somewhere else, there is a chain of intermediary causes and effects linking the two. For example, the vibration of strings in a piano causes your ear to vibrate and you to hear music; in this case a series of vibrations in the air is the link. This principle is unfalsifiable because whenever an apparent counter-example is found the prin-ciple simply requires that some as yet unknown medium exists. This was the case with Newton's theory of gravity, which was always regarded by Newton himself as incomplete precisely because it posited a gravitational force acting between all bodies without explaining how this force was propagated through space. Later, the idea of a field was introduced to solve the problem and this concept was extended to electromagnetic theory which deals with phenomena where similar forces (electrostatic attraction and repulsion) seem to act at a distance, such as the action of a magnet on a compass needle. The pursuit of local theories has certainly been fruitful in the history of science, and the use of other unfalsifiable, and even metaphysical, principles has also had success at various points.

There are also methodological principles that are arguably central to science but not falsifiable. So, for example, many scientists intui-tively regard simple and unifying theories as, all other things being equal, more likely to be true than messy and complex ones. For example, suppose the population of sparrows is noticed to be falling in various regions. Scientists investigating the cause of these separate phenomena will usually seek a unifying explanation, say destruction of hedgerows, which simultaneously explains why sparrows, and perhaps other birds, in different places are all in decline. This prin-ciple is followed in everyday life: if a doctor observes a sudden rise in the number of patients presenting with a particular set of symptoms, he or she will probably assume that a single pathogen is responsible;

if a detective hears reports of a sudden increase in armed robberies in a certain area, he or she will probably look for a single new active gang of robbers; of course, all of them may be wrong but simplicity is only claimed to be one among a number of other *fallible* methodological principles. Some people claim that we have inductive grounds for believing in scientific theories that are simple, unified and so on, because in general the search for simple and unifying explanations has been fairly reliable in producing empirically successful theories, but they would add that we should never make simplicity an absolute requirement because sometimes nature is complex and untidy.

Of course, Popper would reject any talk of our having positive grounds for believing in scientific theories, but the problem for him is that there are many examples of scientists claiming to have been sure they were on the right track when they found a particularly simple or beautiful theory. We ought to apply the requirement of reflective equilibrium to falsificationism just as we did to naïve inductivism, so if it turned out that Popper's theory failed to be compatible with actual scientific practice that would amount to a powerful argument against it. Einstein's special relativity is a wonderful example of a scientific advance that brought unity and simplicity to a messy situation. Often in the case of physics, but also in other sciences, the mathematical formulation of a theory is at the heart of these considerations, and in order to address them properly we need to deal with specific cases closely. However, there is a more fundamental principle of simplicity that is often claimed to be essential to science, namely Occam's razor, which is roughly the prescription not to invoke more entities in order to explain something than is absolutely necessary. (This kind of simplicity is called ontological parsimony.)

We shall discuss the status of these principles in more detail later. For now, note that a falsificationist could argue that it is possible to falsify metaphysical principles by, as it were, proxy. Duhem observed that although a metaphysical theory can never imply a particular scientific theory, it can rule out certain scientific theories. For example, the Cartesian metaphysical picture of a world completely filled with matter, with no empty space whatsoever, is inconsistent with Newtonian mechanics, so arguably the success of the latter counts against the former. This idea could be developed as a response to the present objection to falsificationism, but we shall leave this issue for now.

(d) Hypothesis of natural selection

At one time, Popper was critical of the theory of evolution because he thought the hypothesis that the fittest species survive was tautological, that is to say true by definition, and therefore not falsifiable, yet evolutionary theory is widely thought to be a prime example of a good scientific theory. Most philosophers of biology would argue that the real content of evolutionary theory lies not in the phrase 'the fittest survive', but in the idea of organisms passing on characteristics, subject to mutation and variation, which either increase or decrease the chances of their offspring surviving long enough to reproduce themselves, and so pass on those characteristics. This is supposed to account for the existence of the great diversity of species and their adaptation to the environment, and also to the similarities of form and structure that exist between them. This theory may be indirectly falsifiable but the status of evolutionary explanations is too large a subject for us to enter into here.

(2) Falsificationism is not itself falsifiable

Popper admits this but says that his own theory is not supposed to be because it is a philosophical or logical theory of the scientific method, and not itself a scientific theory, so this objection, although often made, misses its target.

(3) The notion of degree of falsifiability is problematic

The set of potential falsifiers for a universal generalisation is always infinite, so there can be no absolute measure of falsifiability, but only a relative one. Earlier on we discussed the notion of degree of falsifiability where one theory's empirical consequences are a subset of those of another theory. However, often the situation is much more complicated. For example, Einstein's theory of gravitation is supposed to be more falsifiable than Newton's, yet as we have seen empirical consequences can be derived from these theories only if they are conjoined with background theories and assumptions. So we only have reason to believe that high-level and sophisticated theories have the empirical consequences that we think they have to the extent that we believe the background theories and assumptions are themselves likely to be true. The Duhem problem means that judgements about the degree of falsifiability of theories are relative to whole systems of hypotheses, and so our basis for such judgements is past experience and this lets induction in by the back door.

As we will see in the next chapter, this problem becomes more acute if we consider the arguments that some philosophers claim show that all observations are theory laden. If this is correct, then when there is wholesale change in scientific theories there will be a change in what counts as an observable phenomenon and it will be impossible, in general, to compare the empirical content of theories from a point of view that is neutral with respect to them.

(4) *Popper cannot account for our expectations about the future*

In the second quotation in section 3.2 above Popper says that we are not entitled to believe that our best theories are even *probably* true. His position is ultimately extremely sceptical, indeed he goes further than Hume, who says induction cannot be justified but that we cannot help but use it, and argues that scientists should avoid induction altogether. But is this really possible, and is it really plausible to say that we never get positive grounds for believing scientific theories?

Our scientific knowledge does not seem to be purely negative and if it were it would be hard to see why we have such confidence in certain scientifically informed beliefs. After all, it is because doctors believe that penicillin fights bacterial infection that they prescribe it for people showing the relevant symptoms. The belief that certain causes do indeed have certain effects and not that they might not is what informs our actions. For example, according to Popper, there is no positive inductive support for my belief that if I try to leave the top floor of the building by jumping out the window I will fall hard on the ground and injure myself. If observation of past instances really confers no justification on a generalisation then I am just as rational if I believe that when I jump out of the window I will float gently to the ground. I take it that this is an unacceptable consequence of Popper's views for there is nothing more obvious to most of us than that throwing oneself out of high windows when one wishes to reach the ground safely is less rational than taking the stairs. If we adopt Popper's nihilism about induction we have no resources for explaining why people behave the way they do and, furthermore, we are obliged to condemn any positive belief in generalisations as unscientific.

Of course, just when and how we can be justified on the basis of experience in believing general laws and their consequences for the future behaviour of the natural world is the problem of induction. Most philosophers, however, think that solving this problem is not a matter of deciding whether it is more rational to take the stairs but *why* it is more rational to do so. Popper's response to this challenge is to introduce the notion of *corroboration*; a theory is corroborated if it was a bold conjecture that made novel predictions that were not falsified. Popper says that it is rational to suppose that the most corroborated theory is true because we have tried to prove it false in various ways and failed. The most corroborated theory is not one we have any reason to believe to be true, but it is the one we have least reason to think is false, so it is rational to use it in making plans for the future, like leaving the building by the stairs and not by jumping. Popper stresses that the fact that a theory is corroborated only means that it invites further challenges.

However, the notions of boldness and novelty are historically relative; the former means unlikely in the light of background knowledge and therefore highly falsifiable, and novel means previously unknown, or unexpected given existing corroborated theories, so once again induction based on past experience is smuggled into Popper's account. Furthermore, there is an infinite number of best corroborated theories, because whatever our best corroborated theory is, we can construct an infinite number of theories that agree with what it says about the past, but which say something different about what will happen in the future. The theory that gravity always applies to me when I jump into the air except after today is just as corroborated by all my experience up to now as the alternative that tells me not to jump off tall buildings; again we seem to have no choice but to accept the rationality of at least some inductive inferences despite what Popper says.

(5)   Scientists sometimes ignore falsification

If we demand of scientists that they be prepared to state in advance under what conditions they would abandon their most cherished assumptions, then we will be disappointed. We have already discussed the case of the principle of conservation of energy but there are many examples in the history of science where, instead of abandoning a theory, scientists thought up modifications or extra assumptions to save it. Popper admits this but argues that extra assumptions made to save a theory from refutation are acceptable if they entail further predictions. He distinguishes between ad hoc and non-ad hoc modifications of a theory to save it from refutation, and argues that modifications proposed after a falsifying instance must explain the partial empirical success of the old hypothesis, and have further empirical content, otherwise they will be ad hoc and therefore unacceptable within science.

For example, in the nineteenth century Newton's mechanics together with the known facts about the mass, positions and motions of the planets, predicted that the orbit of Uranus should be different from what was actually observed. Instead of regarding their theory as falsified, most scientists of the time assumed that one of the above parameters was wrong, and some proposed the existence of another planet to accommodate the data. This was acceptable according to Popper because this modification increased the empirical content of the science by predicting that this planet ought to be observable. In due course Neptune was indeed observed within one degree of arc of the position that had been predicted, and subsequently this process was repeated as measurements became more precise and Pluto was discovered.

On the other hand, there are certainly extreme cases where most people will agree that a theory has only been saved from refutation by a gratuitous assumption whose only role or justification is to save the theory. For example, in the early twentieth century someone called Velikovsky proposed a theory according to which there had been a series of cataclysms in human history. The theory predicts that there ought to be some record or trace of these events in written or oral history, but no such records are found. This is a clear case of apparent refutation which Velikovsky accommodates by postulating that the cataclysms are so traumatic that collective amnesia prevents people from recording them. This modification is ad hoc because it adds no extra empirical content to the theory. Similarly, if the Bible is literally true, then the Earth is only about six thousand years old and the fossils of dinosaurs, which appear to be much older, seem to refute the biblical theory. However, it is always open to the fundamentalist to argue that the fossils were in fact put in place by God and made to seem much older than six thousand years old in order to test our faith. Both these ways of saving a theory from refutation seem to have a similar structure. The point about them is that there is no independent way of testing the assumption which saves the theory; it merely reconciles the theory with the potentially falsifying evidence.

Unfortunately, it turns out that there are cases in the history of science where a falsifying observation is tolerated for decades despite numerous attempts to account for it. For example, the early atomic theory of Niels Bohr (1885–1962) is actually inconsistent, yet it was widely adopted as a working model. Mercury's orbit was known to be at odds with Newtonian theory for many years yet this never led to the theory being abandoned; finally, Einstein's theory of gravitation predicted the right orbit for the planet and the Newtonian theory was regarded as falsified. It is arguable that Newtonians wouldn't give the conditions under which they would reject the basic assumptions of Newtonian physics, and so it seems lack of commitment is not essential to good science after all. More generally, it often seems to be the case that where scientists have a successful theory, the existence of falsifying observations will not be sufficient to cause the abandonment of the theory in the absence of a better alternative.

3.6 Conclusions

Popper has drawn our attention to features of good science that are now widely emphasised: a critical attitude to the received wisdom, an insistence on empirical content that is precise and wide in scope, and the use of creative thinking to solve problems with bold conjectures that open up radical new possibilities for experiment and observation. The ideas of ad hocness, novel prediction and corroboration must surely play a part in explaining the difference between right and wrong reasoning in science. Lakatos tried to improve upon Popper's falsificationism and avoid some of the problems we have discussed. However, although many scientists insist that theories ought to be falsifiable by experiment, and actively trying to falsify theories may sometimes be important and productive, it seems that we cannot explain the scientific method and the justification of scientific knowledge without recourse to induction of some form or other. Science is about confirmation as well as falsification. At least, that is what many people believe and some of Popper's ideas can help them formulate a more sophisticated inductivism.

The distinction between the context of discovery and the context of justification is used by a sophisticated inductivist to separate the question of how scientific theories are developed from the question of how to test them against their rivals. Sophisticated inductivism is not refuted by those episodes in the history of science where a theory was proposed before the data were on hand to test it let alone suggest it. Instead, the model of hypothetico-deductivism can be adopted. Theories may be produced by any means necessary but then their degree of confirmation is a relationship between them and the evidence and is independent of how they were produced. Since Bacon, there have been many more theories of inductive logic and confirmation including Mill's methods, Whewell's account of consilience, and Carnap's and Reichenbach's mathematical theories of probability. However, in the next chapter we will consider a rather different view of the scientific method.

———o◉o———

Alice: Come on, you can't pretend we never have any positive reason to believe things. I don't know how to justify induction but sometimes it definitely is justified. Do you really think that I have no reason to believe that the next time I catch a train it will be late?

Thomas: I don't know. Maybe we have to form definite beliefs about things to live our lives, but that doesn't mean they are true.

Alice: Well, anyway, science is like everyday life in that respect. If scientists were completely sceptical all the time they wouldn't get anywhere. Sometimes they need to be committed to a theory even if it's got a few problems they can't quite see how to solve.

Thomas: But now I don't really see the difference between science and any other belief system. How can it be okay for scientists to ignore evidence that doesn't suit their prejudices?

Alice: If a theory has lots of other evidence in its favour and it works then it would be crazy to abandon it without something to replace it.

Thomas: Well if it's all a matter of what the competition is like then what we count as so-called scientific knowledge depends on what we happen to have to compare it with, so the same theory could count as knowledge one day and then not the next, just because someone else invented a better theory.

Alice: It doesn't work like that because usually new theories build on old ones so the knowledge in the old theory is preserved as science progresses.

Thomas: But not always. What about when there are revolutions in science?

----•◯•----

Further reading

Falsificationism

Lakatos, I. (1970) 'The methodology of scientific research programmes', in I. Lakatos and A. Musgrave (eds) *Criticism and the Growth of Knowledge*, Cambridge: Cambridge University Press.

Newton-Smith, W. (1981) *The Rationality of Science*, Chapter III, London: Routledge.

Popper, K. (1934, 1959) *The Logic of Scientific Discovery*, London: Hutchinson.

Popper, K. (1969) *Conjectures and Refutations*, London: Routledge and Kegan Paul.

The Duhem problem

Duhem, P. (1906, tr. 1962) *The Aim and Structure of Physical Theory*, New York: Athenum.

Harding, S. (ed.) (1976) *Can Theories be Refuted? Essays on the Duhem–Quine Thesis*, Dordrecht, The Netherlands: D. Reidel.

Lakatos, I. and Musgrave, A. (eds) (1970) *Criticism and the Growth of Knowledge*, Cambridge: Cambridge University Press. See especially the papers by Lakatos and Feyerabend.

Quine, W.v.O. (1953) *From a Logical Point of View*, Cambridge, MA: Harvard University Press. This contains the classic 'Two dogmas of empiricism' but is hard.

4

——•◌•——

Revolutions and rationality

The scientific method is supposed to be rational, and to give us objective knowledge of the world. To say that scientific knowledge is objective means that it is not the product of individual whim, and it deserves to be believed by everyone, regardless of their other beliefs and values. So, for example, if it is an objective fact that smoking causes cancer, or that all metals expand when heated, then it ought to be believed equally by atheists and theists, by conservatives and liberals, and by smokers and non-smokers, if these people are to be rational. Our search for the scientific method has led us from the naïve inductivism of Bacon, which is an account of how to develop scientific theories, to the falsificationism of Popper, which is exclusively concerned with the testing of scientific theories once they have been proposed.

As we saw at the end of the previous chapter, a more sophisticated form of inductivism combines the distinction between the context of discovery and the context of justification, with the view that evidence in science does give us positive reasons for believing both scientific theories, and the generalisations about the future behaviour of things that we can derive from them. Sophisticated inductivism, like falsificationism, departs from naïve inductivism by giving an important role to non-rational factors in the development of science. After all, as we have seen, scientists might be inspired by their religion, their dreams, their metaphysical beliefs or even by blind prejudice when they are developing new theories. For this reason, the context of discovery is outside the domain of rationality; however, the context of justification is subject to the constraints of rationality, and this is supposed to guarantee the objectivity of scientific knowledge.

The main rivals to Popper in philosophy of science in the early to mid twentieth century advocated sophisticated versions of inductivism (often involving mathematical theories of statistics and probability). Indeed, the received view in (anglophone) philosophy of science after the Second World War, which was called *logical empiricism* (a relative and successor of logical positivism), incorporated a commitment to some form of sophisticated inductivism. One of the most important logical empiricists was Carnap, and (following Lakatos 1968: 181) we can concisely express the difference between Hume, Popper and Carnap as follows; Hume thought that science was inductive and irrational, Popper thought it was non-inductive and rational, and Carnap thought it was inductive and rational.

However, in the 1960s, there was a crisis of both realism and rationality in philosophy of science, that has yet to be resolved. There are now many who question the rationality and objectivity of scientific knowledge, largely because of ideas that first came to prominence in the work of a historian and philosopher of science called Thomas Kuhn (1922–1996), who arguably did more than anyone else to cause that crisis in the first place. In contrast to the trio above, Kuhn seems to argue that science is both non-inductive and non-rational. This chapter will mostly be about his account of theory change in science and the philosophical issues it raises, but first it will be helpful to articulate in a bit more detail the received wisdom that he undermined.

4.1 The received view of science

Popper on the one hand, and logical empiricists like Carnap and Reichenbach on the other, disagreed about the correct response to the problem of induction. Popper thought that it showed that confirmation is impossible, while Carnap and Reichenbach thought that it could be solved if the formal details of a precise logic of confirmation could be ironed out. Popper also disagreed with the logical positivists (among whom Carnap and Reichenbach also numbered for a while) about the demarcation of science from non-science. The positivists sought to demarcate the meaningful from the meaningless, by arguing that the meaning of an expression was given by the means by which it could be confirmed. On this view, the meaning of an expression such as 'the temperature of the gas is 100° Celsius' is given exhaustively by a specification of the experimental conditions that would need to occur for someone to justifiably assert the truth of that statement (for example, that if a thermometer was brought into contact with the gas it would give the corresponding reading). We will return to logical positivism in the next chapter. Popper's demarcation criterion does not concern meaning because 'there is a black swan' is perfectly meaningful although it is not falsifiable. However, although there were these important disagreements, there were also many views about the nature of science that Popper, the logical positivists and the logical empiricists shared, including:

(1) Science is *cumulative*. In other words, scientists build on the achievements of their predecessors, and the progress of science is a steady growth in our knowledge of the world. This feature of science is sharply contrasted with other activities, such as art, literature and philosophy, which are progressive in a much looser and controversial sense.

(2) Science is *unified* in the sense that there is a single set of fundamental methods for all the sciences, and in the sense that the natural sciences at least are all ultimately reducible to physics. **Reductionism** is now very controversial but the idea is that, because everything in the world is made of the same basic stuff in complex combinations, the laws of biology ought to be derivable from those of chemistry, and the laws of chemistry from the laws of physics.

(3) There is an epistemologically crucial distinction between the context of discovery and the context of justification. The evidence for scientific knowledge ought to be evaluated without reference to the causal origins of the theories or observations in question; in other words, who made some particular observations and when a theory was proposed and by whom for whatever reason, are irrelevant to the question of the extent to which the observations provide evidence for the theory.

(4) There is an underlying logic of confirmation or falsification

implicit in all scientific evaluations of the evidence for some hypothesis. Such evaluations are *value-free* in the sense of being independent of the personal non-scientific views and allegiances of scientists.

(5) There is a sharp distinction (or demarcation) between scientific theories and other kinds of belief systems

(6) There is a sharp distinction between observational terms and theoretical terms, and also between theoretical statements and those that describe the results of experiments. Observation and experiment is a neutral foundation for scientific knowledge, or at least for the testing of scientific theories.

(7) Scientific terms have fixed and precise meanings.

These theses are also implicit in popular conceptions of the nature of science; however, each of them is apparently at odds with Kuhn's philosophy of science. In the next few sections of this chapter we will look at his views and consider what, if anything, of the image of science presented in (1) to (7) can survive his critique. Along the way we will return to some of the issues that were raised in the previous chapters, and this will further prepare the ground for the discussion of scientific realism to come.

4.2   Kuhn's revolutionary history of science

Kuhn was a physicist who became interested in the history of science and especially the Copernican revolution. The standard view that he found presented in textbooks and in historical and philosophical works, was that the Copernican revolution, and especially the argument between Galileo and the Catholic Church, was a battle between reason and experiment on the one hand, and superstition and religious dogma on the other. Many historians and scientists suggested that Galileo and others had found experimental data that were simply inconsistent with the Aristotelian view of the cosmos. Kuhn realised that the situation was considerably more complex, and he argued that the history of this and other revolutions in science was incompatible with the usual inductivist and falsificationist accounts of the scientific method. Kuhn's book *The Structure of Scientific*

Revolutions (1962) offered a radically different way of thinking about scientific methodology and knowledge, and changed the practice of history of science. His philosophy of science has influenced academia from literary theory to management science, and he seems single-handedly to have caused the widespread use of the word 'paradigm'.

Kuhn argued that many scientists' accounts of the history of their subject considerably simplify and distort the real stories of theory development and change. Often this is because summaries of the evolution of a discipline are intended to motivate and justify contemporary theories, more than they are intended to be faithful to the complexities of history. Kuhn compares the relationship between textbook histories of science and what actually happened to the relationship between a tourist guidebook and what a country and its culture are really like. Obviously guidebooks focus on the aspects of places that the tourism industry want to promote, such as museums and chic café culture, and downplay or omit entirely the aspects that it would prefer to be ignored, such as derelict buildings and hostels for the homeless. Although the story of the Copernican and other scientific revolutions are often told as the triumph of reason and experiment over superstition and myth, Kuhn argued that: 'If these out-of-date beliefs are to be called myths, then myths can be produced by the same sorts of methods and held for the same sorts of reasons that now lead to scientific knowledge' (Kuhn 1962: 2). Kuhn goes on to point out that abandoned beliefs are not thereby unscientific; hence, he argues that the history of science does not consist in the steady accumulation of knowledge, but often involves the wholesale abandonment of past theories.

Already we can see that Kuhn disagrees with (1) and (5) above, but he also suggests even more radical claims about scientific knowledge. As we saw in the last chapter, the Duhem–Quine problem shows us that theory-testing is not as straightforward as is often implied because, when experiment conflicts with a scientific theory, logic alone does not tell us which of the components of the theoretical system is at fault. Although observation and experience certainly constrain scientific beliefs they do not determine them, and hence, argues Kuhn: 'An apparently arbitrary element, compounded of personal and historical accident, is always a formative ingredient of the

beliefs espoused by a given scientific community at a given time'
(Kuhn 1962: 4).

According to Kuhn, the evaluation of theories depends on local
historical circumstances, and his analysis of the relationship between
theory and observation suggests that theories infect data to such an
extent that no way of gathering of observations can ever be theory-
neutral and objective. Hence, the degree of confirmation an experiment gives to a hypothesis is not objective, and there is no single logic
of theory testing that can be used to determine which theory is most
justified by the evidence. He thinks, instead, that scientists' values
help determine, not just how individual scientists develop new theories, but also which theories the scientific community as a whole
regards as justified. This amounts to the denial of theses (2), (3), (5)
and (6) and, as we shall see, he denies (7) as well. In the next section
I will explain the essential details of his philosophy of science.

4.3   Paradigms and normal science

Perhaps the most fundamental concept in Kuhn's philosophy is that
of the scientific *paradigm*. He does not offer us a precise definition of
this term, and indeed it sometimes seems to have a very broad meaning, but we can identify two closely related applications, those of
paradigm as *disciplinary matrix* and paradigm as *exemplar*. Kuhn
argues that before scientific inquiry can even begin in some domain,
the scientific community in question has to agree upon answers to
fundamental questions about, for example: what kinds of things exist
in the universe, how they interact with each other and our senses,
what kinds of questions may legitimately be asked about these things,
what techniques are appropriate for answering those questions, what
counts as evidence for a theory, what questions are central to the
science, what counts as a solution to a problem, what counts as an
explanation of some phenomenon, and so on.

A *disciplinary matrix* is a set of answers to such questions that are
learned by scientists in the course of the education that prepares them
for research, and that provide the framework within which the science operates. It is important that different aspects of the disciplinary
matrix may be more or less explicit, and some parts of it are

constituted by the shared values of scientists, in that they prefer certain types of explanation over others and so on. It is also important that some aspects of it will consist of practical skills and methods that are not necessarily expressible in words. This is partly what makes a paradigm different from a theory, because the disciplinary matrix includes skills that enable scientists to make technological devices work, such as how to focus a telescope, and experimental skills, like how to crystallise a salt from a chemical reaction, which have to be learnt by practical experience (such skills are sometimes called *tacit knowledge*).

Exemplars, on the other hand, are those successful parts of science that all beginning scientists learn, and that provide them with a model for the future development of their subject. Anyone familiar with a modern scientific discipline will recognise that teaching by example plays an important role in the training of scientists. Textbooks are full of standard problems and their solutions, and students are set exercises that require them to adapt the techniques used in the examples to new situations. The idea is that, by repeating this process, eventually, if they have the aptitude for it, students will learn how to apply these techniques to new kinds of problems that nobody has yet managed to solve.

As an example, consider the paradigm of classical, or Newtonian, physics; it consists of at least the following elements:

- background values such as preferences for efficient causal explanations (see Chapter 1), and theories that yield precise quantitative and testable predictions, rather than general and qualitative ones;
- the metaphysical picture of the world as composed of material particles, interacting by colliding with each other, and by attractive and repulsive forces acting in straight lines between particles, and the guiding image of the world as a giant clockwork machine;
- Newton's laws of motion and the law of gravitation, which are the core principles of the paradigm;
- the standard mathematical techniques used to apply the laws to physical systems such as pendulums, collisions of particles, and planetary motions, as well as approximations to account for friction, air resistance and so on;

- the exemplar of Newton's *Principia Mathematica* (the Preface of which explicitly states that Newton's methods will prove applicable in other areas of science).

Important goals of scientists working in this paradigm included extending it to account for electrical and magnetic phenomena and light, and also accounting for the way the gravitational force acted across space in terms of underlying mechanical processes of some kind.

Other examples of paradigms include: Ptolemaic astronomy, the phlogiston theory of combustion (which was based around the idea that combustion is the release of a substance called phlogiston), Daltonian chemistry (which is the chemical theory according to which the elements may be distinguished by their differing atomic weights), the fluid flow theory of electricity (according to which electricity is a material fluid), the caloric theory of heat (according to which heat is a material fluid), particle optics (according to which light is a collection of fast moving, tiny particles), wave optics (according to which light is composed of waves of disturbance in some medium), relativistic physics (according to which the time elapsed between events is relative to the state of motion of an observer, or to be more precise to a frame of reference), and quantum physics (according to which the energy possessed by material objects or electromagnetic waves comes in discrete units, rather than taking a continuous range of values).

Most science is what Kuhn calls 'normal science', because it is conducted within an established paradigm. It involves elaborating and extending the success of the paradigm, for example, by gathering lots of new observations and accommodating them within the accepted theory, and trying to solve minor problems with the paradigm. Hence, normal science is often said to be a 'puzzle-solving' activity, where the rules for solving puzzles are quite strict and determined by the paradigm. Examples of normal science include searching for the chemical structure of familiar compounds, coming up with more detailed predictions and experimental determinations of the paths of planets and other heavenly bodies, mapping the DNA of a particular bacterium, and so on.

According to Kuhn, most of the everyday practice of science is a fairly conservative activity in so far as, during periods of normal science, scientists do not question the fundamental principles of their discipline. Kuhn is very critical of Popper's falsificationism, according to which scientists do and should abandon any refuted theory. It is just not the case, according to Kuhn, that the knowledge of falsifying instances is enough to make most scientists abandon their cherished theories. As I argued in the previous chapter (3.5 number (5)), scientists are often quite committed to their theories, and sometimes they will adopt all manner of strategies to save them from apparent refutation, rather than simply giving them up. If a paradigm is successful and seems able to account for the bulk of the phenomena in its domain, and if scientists are still able to make progress solving problems and extending its empirical applications, then most scientists will just assume that anomalies that seem intractable will eventually be resolved. They won't give up the paradigm just because it conflicts with some of the evidence. Perhaps this is justifiable: after all, if a paradigm has had a great deal of success in the past and has successfully dealt with anomalies that have been found before, then given the massive investment of time and resources that has been made in it, surely it is reasonable to stick with it in the hope that in time the anomaly will be solved. As Kuhn says: 'The scientist who pauses to examine every anomaly he notes will seldom get significant work done' (Kuhn 1962: 82).

However, sometimes scientists become aware of anomalies that won't go away no matter how much effort is put into resolving them. These may take the form of conceptual paradoxes or experimental falsifications. Even these will not necessarily cause much serious questioning of the basic assumptions of the paradigm. However, when a number of serious anomalies accumulate then some, often younger or maverick, scientists will begin to question some of the core assumptions of the paradigm, and perhaps they will begin speculating about alternatives. This amounts to the search for a new paradigm, which is a new way of thinking about the world. If this happens when successful research within the paradigm is beginning to decline, more and more scientists may begin to focus their attention on the anomalies, and the perception that the paradigm is in 'crisis' may begin to take hold of the scientific community.

Crises happen infrequently, according to Kuhn. Paradigms do not get to become established unless they are pretty robust and able to accommodate the bulk of the phenomena in their domains, and it is not easy for a working scientist to question the background assumptions upon which the whole discipline is based. Crises are most likely to occur if the anomalies in question seem directly to affect the most fundamental principles of the paradigm, or the anomalies stand in the way of applications of the paradigm that have particular practical importance, or if the paradigm has been subject to criticism because of the anomalies for a long period of time. If a crisis happens, however, and if a new paradigm is adopted by the scientific community, then a 'revolution' or 'paradigm shift' has occurred. In Kuhn's view, when a revolution occurs the old paradigm is replaced wholesale. So, for example, the adoption or rejection of each of the examples of paradigms listed above is a scientific revolution.

The reader with some knowledge of the history of science will notice that some of the 'revolutions' that Kuhn identified – such as the Copernican revolution – certainly seem to deserve the name, since they involve radical changes in fundamental science, whereas others are more local, and merely involve the rejection of a specific theory within a particular scientific sub-discipline. Nonetheless, Kuhn argues that the structure of these smaller revolutions has much in common with that of the larger and more profound ones. An example of a scientific revolution that is intermediate in the scale of its profundity is the replacement of the phlogiston theory of combustion with the oxidisation theory of combustion. Phlogiston was supposed to be a substance that is released by materials when they burn. Most things, such as wood, lose weight when they are burnt, as this theory requires, but some metals were known to increase their weight when burnt and this was something of an anomaly for the phlogiston theory. However, most chemists in the eighteenth century saw this as no reason to abandon the theory, and it was widely employed by experimentalists who developed various methods for producing different types of 'airs' in their laboratories. Unfortunately, they all used different forms of the theory, and this proliferation of versions of a theory (which the Ptolemaic theory of planetary motions also suffered from in the sixteenth century) is cited by Kuhn as one of the hallmarks of a crisis. Gradually, more and more cases where burning substances results in an increase in weight were identified and, furthermore, the widespread acceptance of Newtonian theory meant that chemists were increasingly inclined to understand mass as a quantity of matter, and hence to think that an increase in mass on combustion must mean that more matter is present after combustion than before it.

Of course, the Duhem problem means that none of this suffices to show that phlogiston does not exist, because many different options to explain these results were possible; for example, some thought that phlogiston might have negative weight, or that as phlogiston left a burning body fire particles might enter and so account for the weight gain. Nonetheless, eventually the phlogiston paradigm was in crisis, and the time was ripe for the acceptance of a new paradigm. This was instigated by the chemist Antoine Lavoisier, who proposed (in 1777) that phlogiston does not exist, and that combustion involves, not the loss of phlogiston, but the gain of oxygen. This revolution saw a specific theory in chemistry being replaced, but also a fundamental change in the methods that were considered appropriate in chemical experiments. Up until this time it had been thought almost universally that there was really only one kind of 'air', although it could be of different degrees of purity; after this revolution it was accepted that oxygen is but one gas among the constituents of ordinary air.

There are two points about Kuhn's account of this and other scientific revolutions that must be emphasised:

- This is a completely different view of scientific change to the traditional idea of cumulative growth of knowledge, because paradigm shifts or scientific revolutions involve change in scientific theories that is not piecemeal but holistic. In other words, the paradigm does not change by parts of it being changed bit by bit, but rather by a wholesale shift to a new way of thinking about the world, and this will usually mean a new way of practising science as well including new experimental techniques and so on.
- Revolutions only happen when a viable new paradigm is available, and also when there happen to be individual scientists who are able to articulate the new picture to their colleagues.

It is ironic that, in a way, Popper's account of the history of science gives a more central role to revolutions than Kuhn's, because, for the former, science is in a state of permanent revolution where fundamental principles are constantly being subjected to test, and where criticism is ubiquitous and relentless. For Kuhn, on the other hand, revolutions are a pretty rare occurrence and most science is normal science, where fundamental principles are not called into question, and the ongoing work of scientists is fairly routine. Popper thinks we can reconstruct the history of science as a series of rational decisions between competing theories based on experimental evidence. On the other hand, Kuhn thinks that, because revolutions involve a change in the context within which scientific questions are normally resolved, the evidence alone can never be enough to compel scientists to chose one paradigm over another. After a revolution scientists have a new way of looking at things and new problems to work on, and old problems may be simply forgotten or regarded as irrelevant rather than being solved. (Hence, the idea that the empirical content of successor theories builds upon that of their predecessors, which both Popper and the positivists maintain, is false according to Kuhn.) The heart of the disagreement between them is that, whereas Popper thinks that commitment to theories is anathema to the scientific enterprise, Kuhn emphasises that most scientists most of the time are thoroughly committed to the paradigm they are working within, and in the face of refuting evidence they are quite unlikely to locate the problem within the central assumptions that define the paradigm. Only when there is a crisis will scientists consider replacing the paradigm itself, and when this is happening there is not really science being done at all.

Kuhn argues that scientists' values play an important role in determining whether they accept a new paradigm or not. For example, Einstein in his mature years believed strongly that science ought to provide an account of how the world is, rather than just giving us empirically adequate theories; in other words, he was a scientific realist. On the other hand, some of the founders of quantum mechanics thought that the aim of a physical theory is only to provide a means of predicting the phenomena; in other words, they were instrumentalists. It turned out that quantum mechanics was developed, and it was soon found to be highly predictively successful, but even today, decades later, there is no widely agreed upon way to interpret the theory realistically. Hence, Einstein never accepted quantum mechanics, while many other scientists did, and the dispute between them is not about the empirical evidence in support of the theory, but about what is to be valued in scientific theories. (Einstein's own theories were derided as 'Jewish physics' by some scientists in the early part of the twentieth century.)

Kuhn also emphasises the role of psychological and sociological factors in disposing scientists to adopt or a reject a particular paradigm. Some people are inherently more conservative than others, while some enjoy being a lone voice in the wilderness; some people are risk takers and others are risk averse, and so on. Obviously, a scientist at the end of his or her career who already holds a professorship has more freedom to indulge in speculations on the fringes of the subject than a young researcher on a temporary contract. Every scientist is also influenced in how they see the world by who happens to be their teachers and students. So paradigms are the intellectual property of social groups whose rules and conventions are to be found, not just in their textbooks and theories, but also in the nature of funding bodies, research and educational institutions, the review boards for learned journals and so on. In Kuhn's view, science must be seen in its social and historical context, and this means that scientific change cannot be properly understood without taking account of social forces. If this is right, the commitment within the received view to a purely logical account of the relationship between theories and the evidence for them, and hence to an objective measure of the justification of scientific theories by observational data (4.1(4) earlier) is entirely misconceived. It seems that, as Lakatos puts it: 'according to Kuhn scientific change – from one "paradigm" to another – is a mystical conversion which is not and cannot be governed by the rules of reason: it falls totally within the realm of *(social) psychology of discovery*' (Lakatos 1968: 151).

4.4   The Copernican revolution

In Chapter 1, I explained some of the story of the Copernican revolution. It seems to have inspired many of Kuhn's ideas, so it is appropriate to illustrate these ideas with reference to this revolution. Although Galileo's celebrated advocacy of the Copernican theory against the Catholic Church happened in the early seventeenth century, the whole process, during which the Ptolemaic geocentric paradigm was abandoned in favour of heliocentrism, took around 150 years. Eventually, by the end of the seventeenth century, Newton's theory of gravity offered a unified account of the motions of the planets, the Moon's influence on the tides, and much more besides. However, although with hindsight we can see that the resulting worldview is more complete, unified and empirically adequate than the one it replaced, this was not part of the evidence available to those who instigated the shift to heliocentrism.

The Ptolemaic paradigm had many things in its favour. For example, the cosmology with the Earth at the centre of the universe seemed natural to those who believed that God created it specifically for human beings, and since we do not feel the Earth to be in motion, it seems appropriate to believe it to be the fixed centre around which everything else rotates. Furthermore, this picture allowed theologians to locate heaven literally above the Earth, and Aristotle's account of the natural motions of the heavenly bodies gave a neat explanation of what is observed in the night sky. Ptolemy's basic theory offered a reasonably accurate means of predicting the motions of the planets and was used successfully for centuries.

However, the paradigm faced certain anomalies because the orbits of the planets did not seem to be perfect circles, but it was possible to adapt the theory of planetary motions with the introduction of epicycles and eccentric orbits and so on (as explained in Chapter 1). This perfectly fits the picture of normal science: astronomers gather more detailed data, and where it doesn't fit the paradigm, rather than rejecting its basic assumptions, they find ingenious ways to solve problems and accommodate the known phenomena. The Ptolemaic paradigm became successively more complex, but anomalies remained and provided an ongoing research programme. Eventually these anomalies built up. The complexity and number of the competing versions of the Ptolemaic theory proposed to accommodate them, and the social pressure for calendar reform, which made resolving the anomalies and formulating a definitive theory a priority, led to a widespread perception that the paradigm was in crisis. Ultimately, the revolution did happen but there were two further conditions that seem to have been necessary. The first was the existence of an alternative theory, and this is what Copernicus provided. However, that alone would not have been sufficient unless there had been individuals who were keen to work on the new paradigm such as Kepler, Galileo, Descartes and others.

It is here that we might begin to suspect that this revolution had a non-rational nature, because each of these thinkers was motivated by quite different reasons to adopt the Copernican picture. Since each of them chose to adopt the Copernican paradigm when it was not yet fully developed, and when it faced many unsolved problems, they took a considerable intellectual risk. None of them could be sure that it would ultimately provide a more adequate account of what we observe in the night sky, and indeed initially Copernicus' theory was no more accurate than its Ptolemaic predecessor. The evidence on either side was never conclusive, and there was much that the old paradigm could account for better than the new one. After all, the new paradigm was completely at odds with background beliefs about the place of human beings at the centre of the cosmos, and it also contradicted the best physical theory people had, namely Aristotle's. Furthermore, Copernicus' theory implied that sometimes the Earth is on the same side of the Sun as the planets Venus and Mars, and sometimes it is on the opposite side. According to the theory, given the distances involved, Venus ought to appear to be up to six times as large at some times as at others. However, observations with the naked eye were unable to detect any change in size. Later, the difference in size was observed with telescopes, but the fact remains that at the time when Copernicus proposed his theory it was inconsistent with the observational evidence.

Later, Brahé (the astronomer whose instruments gathered the data that Kepler studied before formulating his laws of planetary motion) derived another prediction from Copernicus' theory which was then falsified by observation. He argued that if the Earth moves, then the direction in which the distant stars are observed from the Earth ought to vary as the Earth passes from one side of the Sun to the other. He tried to detect this effect, which is known as stellar parallax, with his instruments, and failed. Since his instruments were the most accurate available at the time he concluded that the Copernican theory was false.

Copernicus' theory also faced formidable arguments that seemed to refute it. One of the most compelling was the 'tower argument', which goes as follows: consider what should happen if the Earth is moving and if a stone is dropped from a tall tower. The base of the tower will move some distance while the stone is falling, so the stone ought to land some distance away from the base of the tower. Yet if such an experiment is performed the stone is observed to land the same distance from the bottom of the tower as it was from the top of the tower when it was released. Hence the Earth cannot be in motion. Similarly, if the Earth is moving, why aren't the objects on the surface of the Earth thrown off as grains of sand placed on the rim of a wheel are thrown off when the wheel is spun? Another consideration was that there was no explanation within the Copernican theory for why the Earth does not leave the Moon behind as it orbits the Sun. (This was why Galileo's observation of the moons of Jupiter (in 1609) was so important, because Galileo's geocentric opponents believed that Jupiter moved, and he could argue that if Jupiter could orbit without losing its moons then so too could the Earth.)

All of these arguments were known to advocates of heliocentrism, and yet none could be satisfactorily answered during the early stages of the Copernican revolution. So, as well as solving some problems, the new theory raised all manner of new ones. Depending on their values, different people reacted differently: for example, those who prized mathematical simplicity had a very good reason to adopt the Copernican paradigm, while those who prized coherence of the over-all world-picture and conformity to common sense (we don't feel the Earth moving) were motivated to stay with the Ptolemaic one. Kuhn argues that it is quite implausible to think that they all carefully weighed up the evidence and then chose between the alternative paradigms on compelling rational grounds. The characters of individuals and their beliefs gave each of them different reasons for making their contributions to the Copernican revolution. It seems that what counts as a rational ground, and the relative weights that ought to be given to different rational grounds are negotiable.

In the case of Copernicus, something about his character made him prepared to revise radically the mathematical description of the solar system, rather than being content with adjusting the Ptolemaic system. He also happened to have the mathematical ability to formulate the alternative heliocentric system precisely. Galileo happened to be belligerent and rebellious enough to take the Church on, even when the consequences for his own life were unpleasant. Kepler is said to have had a mystical belief in the basic mathematical harmony of the natural world that made him prefer to put the planets in simple elliptical orbits rather than complex circular ones, and he also had access

Similarly, we might naturally assume that certain phenomena are associated with motion and not with being at rest, because in our experience things moving and things at rest are quite distinct. However, in modern physics uniform motion and rest are not physically different, indeed the difference between them is entirely relative to one's frame of reference. It is clear both that some division of phenomena into types is required before science even begins, and that such taxonomies may be revised when new theories are adopted. It is also clear that, once a science has matured, the idea that further observations should be presuppositionless is undesirable, because it means starting from scratch instead of building on previous success.

However, although existing theories guide us in developing new theories, and tell us which observations are significant and so on, the distinction between the context of discovery and the context of justification can be invoked to maintain the idea that scientific theories are tested by observations. Many empiricist philosophers have drawn a sharp distinction between the observational and the theoretical, and both logical positivists and Popper, at least in his earlier work, assume it. According to the received view, the theory-independence or neutrality of observable facts makes them a suitable foundation for scientific knowledge, or at least for testing theories (see 4.1(4) above). The received view incorporated a distinction between observational terms, such as 'red', 'heavy' and 'wet', and theoretical terms, such as 'electron', 'charge' and 'gravity'. The idea is that the rules for the correct application of observational terms refer only to what a normal human observer perceives in certain circumstances, and that they are entirely independent of theory. So, for example, Ernest Nagel (1901–1985) in his influential book *The Structure of Science* (Nagel 1961) argues that every observational term is associated with at least one overt procedure for applying the term to some observationally identifiable property, when certain specified circumstances are realised. So, for instance, the property of being red is applied to an object when it looks red to a normally functioning observer in normal lighting conditions. Many other writers analyse the logic of theory testing relying upon this distinction between observational and theoretical terms.

On the other hand, Kuhn was one of those who emphasised what has become known as the theory-laden nature of observation. The idea here is summed up by the philosopher N.R. Hanson

(1924–1967), who said: 'Seeing is not only the having of a visual experience; it is also the way in which the visual experience is had' (Hanson 1958: 15). He argued that the visual experience of two observers may be different, even when the images on their retinas are identical. He thought this because he thought that interpretation cannot be separated from seeing. In general, according to Hanson, 'observation of x is shaped by prior knowledge of x' (Hanson 1958: 19). Some famous examples show that sometimes the nature of our perceptual experience depends on our previous experience and concepts (see Figure 2).

The idea is that we can see the cube either with the top right square facing us at the front of a cube, or the bottom left square as the front of it. The other picture is a rabbit's head facing left or a duck's head facing right. There are several things to notice about the experience of looking at these pictures. First, it is only because we are used to seeing three-dimensional objects represented in a two-dimensional picture that we see them as pictures of anything at all. Secondly, we can learn to see the pictures differently, and so how we experience them is not fixed. Most people find one or other way of seeing each picture more natural, and initially have to make an effort to bring about the other experience. Thirdly, when one's experience changes from say seeing a duck to seeing a rabbit, the change is a 'gestalt shift', in other words, the character of the experience as a whole changes.

An example of scientific observation that is often discussed is that

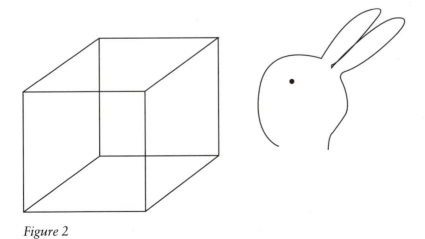

Figure 2

of looking through a microscope; whether something in the image is classed as a real object, or as merely an artefact of the staining process used to prepare slides for examination, is determined by background theory. In general, it is often the case that scientists have to learn to make observations with particular devices. Hence, the experienced doctor looking at an X-ray photograph of a fracture can 'see' all kinds of details that are quite invisible to the layperson. Kuhn and others used such examples to argue that, in general, what scientists perceive is partly determined by their beliefs; a Copernican looking at the sunset sees the Sun stay still and the horizon rise, while a Ptolemaic astronomer sees the horizon stay still as the Sun falls behind it. This threatens to undermine the objectivity of scientific theory testing, because if all observations are contaminated by theories, then observation cannot be the neutral arbiter between competing theories that the received view says it is. If this is right then the history of science is likely to involve various cases where the gathering of observational evidence is biased by the presuppositions of the observers.

An example where this seems to happen is the case of sunspots (which are patches of a different brightness on the surface of the Sun) which were never recorded in Europe prior to the Copernican revolution, but which were well known to Chinese astronomers for centuries beforehand. It seems that the Europeans' belief that the heavens were a perfect unchanging realm prevented them from seeing this obvious example of a changeable extraterrestrial phenomenon. Another case, which is cited by Hanson, is the failure of physicists to notice the tracks in cloud chambers caused by positrons before the theoretical postulation of these particles (in 1928) by Paul Dirac (1902–1984). When particle physicists look back at the experiments conducted in the years before Dirac's work they see clear evidence of positrons that seems to have been completely missed by their predecessors.

There are other cases where what is observed, or regarded as observable, seems to be contaminated by theory. According to Galileo's theory of motion, only relative motion is observable and hence we do not feel the movement of the Earth relative to the Sun, because it isn't moving relative to us. For his critics, all motion is observable, and so the fact that the Earth does not move is established by everyone's experience. Galileo also argued with Aristotelian philosophers about what could be observed with the telescope. For him, when trained on Jupiter, it clearly showed that the planet had its own moons, yet his critics were sceptical about the reliability of this new instrument. Although modern astronomers agree with Galileo about the moons of Jupiter, it now seems he mistook Saturn's rings for moons, and also that, in some cases, he mistook optical illusions for craters on our own Moon.

A further argument for the theory-ladenness of observation appeals to the fact that there seems to be a continuum between cases where someone observes something and cases where someone infers something. For example, suppose someone claims to observes a jet plane in the sky; doesn't he or she actually observe a dot and a vapour trail and infer that it is a plane? Similarly, a scientist is said to observe an electric current in a wire by the movement of a needle on an ammeter, or the lighting of a bulb; but in either case couldn't he or she be said to have inferred the presence of the current rather than to have directly observed it? Some entities such as trees and birds are clearly observable, but what of bacteria, molecules and electromagnetic waves? It is also worth pointing out that we can sometimes use theoretical language to describe observable objects; for example, we talk about 'lighting the gas', 'microwave ovens', 'silicon chips', and so on. It seems that much of our ordinary language is theory-laden, and also that we often describe cases where, strictly speaking, we inferred the presence of something as if we had directly observed it.

However, it is important not to confuse the idea that all language used to describe observation is theory-laden, and the idea that the observations themselves are theory-laden. It is quite plausible to argue that, in language, the line between observational and theoretical terms is blurred, but much more controversial to argue that what theories we believe actually affects the content of our observations, rather than just what we pay attention to and how we describe it. The philosopher Paul Churchland believes that perception is 'plastic' in the sense that the nature and content of our sensory perception is affected by what theories we use to think about and describe the world: 'we learn, from others, to perceive the world as everyone else perceives it' (Churchland 1979: 7). He argues that over a period of time the way we perceive the world may change quite drastically if we come to believe new theories.

An opposing view is put by Jerry Fodor: 'given the same stimulations, two organisms with the same sensory/perceptual psychology will quite generally observe the same things' (Fodor 1984: 24). Fodor maintains that some beliefs are directly fixed by observation, which is to say by the activation of the senses, and he distinguishes these from beliefs fixed by inference. Those who, like him, oppose the more radical conclusions of Churchland and Kuhn appeal to a distinction between 'seeing that . . .' and merely 'seeing . . .'. Of course, someone without the relevant concepts cannot see *that* there is a glass of water in front of them, but they can see the glass of water, as may be evident from the fact that they pick it up in curiosity. So it may be argued that in science to see *that* there is a planet in a certain part of the sky requires theorising it as such, but simply to see a speck of light in that part of the sky merely requires having a properly functioning visual system.

I hope the simple examples I have discussed make the idea of the theory-ladenness of observation vivid, but of course a proper examination of this issue requires familiarity with the work of psychologists and cognitive scientists on human perception. In fact, there are a host of experiments which show that, at least in some situations, what people perceive depends to some extent on their concepts and beliefs. On the other hand, we do not always see what we expect to see, and there is some evidence against the idea that observations are contaminated by our beliefs and concepts. For example, consider the Müller–Lyer illusion (see Figure 3).

Even when we measure the lines carefully and determine that they are the same length, we still perceive one to be longer than the other. Here, then, is a case where it seems our perception is quite immune from contamination by our beliefs.

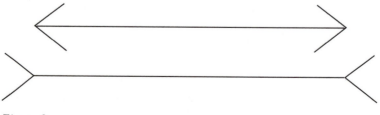

Figure 3

At one stage, the positivists thought that observation statements were the certain foundations of science, but if they are theory laden then they are only as certain as the theories they presuppose. However, although observation reports may not be neutral with respect to all theories, they may still be neutral with respect to the theories they are being used to decide between. Anyway, the general case for saying that the content of observations themselves is a function of the theories the observer believes cannot be established merely by the sorts of example shown above, or by a few cases from the history of science where scientists have disagreed about what is observed on a given occasion. Clearly, the relationship between observation and theory in science is one that deserves detailed investigation. I hope I have at least awakened the reader to many interesting questions about it, showed that the naïve view of a sharp distinction between theoretical and observational terms is implausible, and also made clear the distinction between different things that may be meant by saying that observation is theory-laden. Of course, what people notice or choose to report, and how they report it, is influenced by the theories they have about the world. While our investigations have shown that it certainly is true that theory guides observation, and it is probably true that no interesting observational descriptions are theory-neutral, the case for saying that what is actually seen is different depending on what theory is held is not easily proven.

4.6 Incommensurability

Incommensurability is a term from mathematics which means 'lack of common measure'. It was adopted by Kuhn and another philosopher, called Paul Feyerabend (1924–1994), both of whom argued that successive scientific theories are often incommensurable with each other in the sense that there is no neutral way of comparing their merits. One of the most radical ideas to emerge from Kuhn's work is that what counts as the evidence in a given domain may depend upon the background paradigm. If this is right, then how can it be possible rationally to compare competing paradigms? Kuhn's argues that there is no higher standard for comparing theories than the assent of the relevant community, and that '[the choice] between competing paradigms proves to be a choice between incompatible modes of community life' (Kuhn 1962: 94).

Hence, he can be read as suggesting that, far from being based on the evidence, so-called scientific progress is driven by nothing more than mob psychology, and that the empirical confirmation of hypotheses is a rhetorical sham. (This inspired what is known as 'the strong programme in the sociology of knowledge', which aims to explain scientific theory change in terms of psychological and social forces.) Many people have used Kuhn's arguments to support what philosophers call relativism about scientific knowledge, which is the view that the 'truths' of scientific theories are determined in whole or in part by social forces. A simple form of **epistemic relativism** would say that, for example, a particular theory in physics or biology might be counted as knowledge just because it was believed by those with status and influence within the community of physicists or biologists.

The idea that competing paradigms are incommensurable is supported by the theory-ladenness of observation; if it is true that all observations are contaminated by background theories then the merits of each paradigm cannot be compared by subjecting them to experimental test because the proponents of the competing paradigms will not necessarily agree about what is observed. We have seen that this was the case in the arguments between Galileo and the Church about whether the Earth moves or not. The Copernican Revolution is an example of how, when a paradigm changes, so too do the methods that are appropriate for testing certain theoretical principles, and so too do the problems that science has to solve. For the modern scientist, a body will remain at rest or in uniform motion unless a force acts to change this state, hence there is no need to explain what keeps, say, an arrow in flight once it is accelerated by the bowstring, rather the problem is to explain how gravity and air resistance combine to prevent it continuing in a straight line forever. For the Aristotelian, on the other hand, there is a pressing need to explain what keeps an arrow in the unnatural state of motion after it has left the bow.

Clearly, different people classify things in the world in sometimes radically different ways. On occasion it seems that to evaluate people's beliefs we must understand particular assertions they make in relation to their whole linguistic practice. Some earlier work in science may sometimes be comprehensible in the light of later theories; for example, the caloric theory of heat developed by Pierre Laplace (1749–1827), according to which heat is a material substance, allowed Laplace to calculate the speed of sound in air very accurately. A contemporary physicist can quite readily understand his methods despite the fact that heat is now regarded as a form of energy associated with the vibration of molecules. On the other hand, the reasoning of a Renaissance thinker like Paracelsus (1493–1541) is almost incomprehensible to the modern scientist because his whole way of looking at the world and the types of answers he is seeking are completely alien to a modern scientific outlook. For example, he argues that a plant whose leaves have a pattern that looks like a snake will provide protection against poisons, and that good doctors should not have a red beard. Indeed, some of his statements do not seem to be merely false but rather not even candidates for truth or falsity, because they owe their intelligibility to forgotten styles of reasoning. Hence, questions from old paradigms are not always answered because sometimes we come to think that they don't even make sense.

Kuhn likens a paradigm change to a 'gestalt switch' of the kind one experiences when alternately seeing the picture above as a duck and then a rabbit. The point about gestalt switches is that they are holistic. Similarly, the differences between paradigms at the level of concepts, ontology and so on are global and systematic. Theories within different paradigms are incommensurable, in the sense that the terms and concepts of scientific theories in different paradigms are not mutually intertranslatable; this is called *meaning incommensurability*. Kuhn assumes that scientific terms get their meaning from their position in the structure of a whole theory. For example, 'mass' in Newtonian theory means something different from 'mass' in Einstein's relativity theory. It seems then that when we compare the status of a sentence featuring the term mass in these two theories, we are really comparing two sentences with different meanings. In the Copernican revolution, the idea of motion underwent a radical change. Can we really say that the Aristotelians and Galileo have different theories about the nature of motion, or should we say that they simply mean different things by the word 'motion'? There is no definite answer to this question for Kuhn because scientific terms do not always have fixed and precise meanings (hence he denies 4.1(7) above).

It was widespread before Kuhn's work for philosophers to believe that what a particular scientific term, say 'atom', refers to is determined by what the theory says about atoms. If this is right then different theories about 'atoms', which say different things about them, will actually refer to different things. This is called *reference incommensurability*, and it is bad news for realism, for it suggests that different theories about 'electrons' are actually all about different things, and hence there is no reason to believe that science has made progress in understanding the underlying nature of things. This seems to imply that there is no one way the world is, but that rather the world we live in is an artefact of our theories about it. Indeed, Kuhn says that 'when paradigms change the world changes with them' (Kuhn 1962: 111). On this view, the different languages of different theories correspond to the different worlds of different theories, and the proponents of competing paradigms inhabit different worlds; for example, the world of Einstein is literally a different world from that of Newton. Consequently, we cannot say that Copernicus discovered that Ptolemy and earlier philosophers were wrong to think that the Earth revolves around the Sun, because Copernicus' Earth is literally a different object from Ptolemy's. In this way, Kuhn has been perceived as undermining the notion of scientific truth and even of an objective reality. Hence, there are some people who argue not that scientific knowledge is relative, but that reality itself is socially constructed. So, for example, it is sometimes said that physicists literally construct electrons in their laboratories. On this view, which is called **social constructivism**, an electron has the same ontological status as say a political party, or a nation state, in the sense that both only exist because people believe they exist.

4.7 Relativism and the role of reason in science

Since Kuhn's work, there has been an intense debate about many of the issues he raises. The stakes are high in any debate about science because, as I have emphasised, what we regard as scientific has such an effect on our lives. It is not clear what exactly the sides are in the so-called science wars, but unfortunately, as with so many issues, we can approach this one by considering the extreme positions on each side. On the one hand are those who hold science up as the source of all knowledge, and the only intellectually legitimate form of inquiry. According to them, not only are the teachings of the book of Genesis scientifically proven to be wrong, but we have no need of the myths of any culture, because modern science gives us a comprehensive account of most natural phenomena and the history and geography of the Earth and the entire universe. Of course, scientists vary in their evangelism, but in any bookshop one can find texts offering grand scientific explanations of language, the mind, ethics, human behaviour, the creation of the universe and so on. The most extreme defenders of science think their opponents are superstitious and irrational. On the other hand, there are those who argue that there is nothing special about science, and that indeed it may be worse, or at least no better, than creation myths.

However, it is quite possible to defend the rationality of science without being committed to reductionism about the mind, atheism, the invalidity of other forms of inquiry and so on. Furthermore, a defender of scientific rationality might at the same time be highly critical of the contemporary practice of some or all of the sciences, and highly sceptical about some particular scientific theories. Someone who has a definite account of when science is being conducted properly is able to criticise a particular scientific community on a principled basis. For example, it is plausible that the free exchange of ideas and information is an essential feature of good science. Hence, if the commercial interests of their sponsors are interfering with scientists' freedom to communicate, this can be criticised as unscientific.

Kuhn's history of various scientific revolutions shows us that individual scientists do not live up to the philosopher's ideal of maximally rational agents, always making decisions based on the evidence independently of their own personal interests and goals. On the contrary, according to Kuhn, scientists are often very much attached to a paradigm, and sometimes particular individuals will do almost anything to retain it in the face of contradictory evidence, including perhaps, distorting experimental data, using institutional power to stifle dissent, using poor reasoning and bad arguments to defend the status quo, and so on. Indeed, sometimes the established scientists will refuse to adopt the new paradigm and, rather than being persuaded by rational argument, eventually they simply die out, while the next generation get on with developing the new approach. Of course, disreputable behaviour and fallacious reasoning seem to be features of all spheres of human life, so it would be pretty surprising if they were never found in science, and clearly the idea that all scientists are saint-like pursuers of the truth is unrealistic to the point of being ridiculous. Kuhn pointed out that much of the practice of science is relatively routine, requiring a great deal of technical knowledge, but not necessarily a great deal of critical thought.

Most philosophers of science now accept that their theories about science need to be informed by detailed historical work that is sensitive to the context in which the scientific theories of the past were developed, and that it is usually advisable not to take the textbook accounts of the history of science at face value. Far greater attention is now paid to what scientists actually do rather than just to what they say they do. Kuhn's deconstruction of traditional ideas about the scientific method and the relationship between theory and observation has inspired a host of researchers in the history, sociology and philosophy of science and technology to undertake careful study of the practice of science, including experimental techniques and so on, where before practice was often ignored in favour of attention to theory.

In the light of the preceding discussion some scepticism about scientific knowledge seems inevitable. Not even the most realist and rationalist philosophers of science would argue that all established scientific theories are proven to be true beyond any doubt, nor even that they are *all* probably true. But how far should a healthy scepticism go? The worry about normal science is that whatever anomalies exist already are really a crisis waiting to happen. If this is right then why should we believe in our best current theories? Kuhn has been accused of being a relativist and of being a constructivist, and even those who do not ascribe these views to him have accused him of inspiring them in others. However, Kuhn has sought to clarify his ideas and to show that he was never as extreme as people have thought. It ought to be clear by now why many philosophers have interpreted Kuhn as arguing that changes in scientific theories are partly determined by social and psychological factors, rather than being purely a matter of rational appraisal of the evidence. However, this cannot be the whole story because revolutions in scientific thought do happen, even when they are very inconvenient for the scientific establishment – by requiring textbooks to be rewritten and so on. Furthermore, it is hard to believe that theory acceptance in science is merely a matter of whim, prejudice and so on, in the light of the incredible success of scientific theories and their applications to technology.

In his later work, Kuhn sought to distance himself from extreme views which give no role to rationality in the progress of science, and which do not allow for comparison of the merits of theories within different paradigms. He argues that the following five core values are common to all paradigms:

- A theory should be empirically *accurate* within its domain.
- A theory should be *consistent* with other accepted theories.
- A theory should be wide in *scope* and not just accommodate the facts it was designed to explain.
- A theory should be as *simple* as possible.
- A theory should be *fruitful* in the sense of providing a framework for ongoing research.

Hence, Kuhn avoids complete irrationalism because these values impose some limits on what theories scientists can rationally accept. On the other hand, these values are not sufficient to determine what decisions they ought to make in most interesting cases, because these values may conflict; a theory may be simple but not accurate, or fruitful but not wide in scope, and so on. Furthermore, a value like simplicity may be understood in different ways depending on background views and so on.

Whether or not some role for rationality in theory change is compatible with Kuhn's philosophy of science, it is clear that his account threatens to undermine each of the seven aspects of the traditional view of science with which I began this chapter. Science is not cumulative because paradigm shifts involve the abandonment of old theories rather than the steady accumulation of knowledge; science is not unified because everything within a sub-branch of science is relative to the dominant paradigm which is, in general, not shared between different sciences. There is no neutral standpoint from which to appraise theories so the context of justification is an illusion, as is a single logic of testing theories since all judgements of the worth of a theory are made from within a paradigm. Science is not value-free because social and psychological factors play an ineliminable role in theory choice and there is therefore no sharp distinction between scientific theories and other belief-systems. The only way we can demarcate science from non-science is by appealing to the puzzle-solving nature of normal science, and Kuhn's five core scientific values, but the ranking of the latter is not determinate and so they have little analytical force.

For those who are unconvinced by the more radical claims of Kuhnian philosophy of science, the problem of explicating the nature of the scientific method remains. In the second half of this book questions about the scientific method will continue to lurk in the background, but the primary focus will be on the contemporary debate about whether we ought to believe, not just in the empirical generalisations that science incorporates, but also in the unobservable entities and processes that much of modern science describes.

—◦◉◦—

Alice:	I'm not denying that science doesn't change radically sometimes, but it isn't that often and I still think that the theories we have now are improvements on past ones.
Thomas:	Yes, but don't you see – when scientists are evaluating theories sometimes all their background beliefs and values affect their judgements. Society gets scientific theories that reflect its other beliefs.
Alice:	Look, of course social factors make a difference to science in the short term, but in the long run it's going to be the true theories that win out.
Thomas:	But our decisions about what theories are true aren't just mechanically decided by doing experiments. We have to see how they fit in with the rest of science, which makes the whole business of evaluating theories relative.
Alice:	Up to a point maybe, but in the end a theory will either work or not and that's the real test. I mean the reason people believe in atoms and molecules and stuff is because

that has helped us build computers and design new drugs and send rockets to the Moon. You can't deny the power of science. People with all sorts of values use the theories of modern science because they work.

Thomas: So that's what's left of your scientific method is it? What works must be right. Where does that get us with the Big Bang then? I am not aware of that theory being much use so far.

Alice: I guess I don't know what the scientific method is really about. Perhaps it's more to do with scientists as a whole debating ideas and sharing information, rather than each individual scientist being completely open-minded and following a procedure. But going back to the Big Bang, I believe in it because it fits in with our best theories and it explains and predicts what we observe with telescopes and other devices.

Thomas: But theories don't have to be to true to describe what we observe correctly.

—·◦☉◦·—

Further reading

Kuhn's philosophy of science

Hacking, I. (1983) *Representing and Intervening*, Introduction and Chapter 5, Cambridge: Cambridge University Press.

Hoyningen-Huene, P. (1993) *Reconstructing Scientific Revolutions: Thomas Kuhn's Philosophy of Science*, Chicago: University of Chicago Press.

Kuhn, T.S. (1962, 2nd edn 1970) *The Structure of Scientific Revolutions*, Chicago: University of Chicago Press.

Kuhn, T.S. (1977) *The Essential Tension*, Chicago: University of Chicago Press.

Lakatos, I. and Musgrave, A. (eds) (1970) *Criticism and the Growth of Knowledge*, Cambridge: Cambridge University Press.

On the Copernican revolution

Feyerabend, P. (1977) *Against Method*, London: New Left Books.

Kuhn, T.S. (1957) *The Copernican Revolution: Planetary Astronomy in the Development of Western Thought*, Cambridge, MA: Harvard University Press.

The theory-ladenness of observation

Churchland, P. (1979) *Scientific Realism and the Plasticity of Mind*, Cambridge: Cambridge University Press.

Couvalis, G. (1997) *The Philosophy of Science: Science and Objectivity*, Chapter 1, London: Sage.

Feyerabend, P. (1977) *Against Method*, Chapters 6–11, London: New Left Books.

Hacking, I. (1983) *Representing and Intervening*, Introduction and Chapter 5, Cambridge: Cambridge University Press.

Hanson, N.R. (1958) *Patterns of Discovery*, Cambridge: Cambridge University Press.

Incommensurability

Hacking, I. (1983) *Representing and Intervening*, Chapter 6, Cambridge: Cambridge University Press.

Papineau, D. (1979) *Theory and Meaning*, Oxford: Oxford University Press.

Shapere, D. (1981) 'Meaning and scientific change', in I. Hacking (ed.) *Scientific Revolutions*, Oxford: Oxford University Press.

The rationality of science

Feyerabend, P. (1977) *Against Method*, London: New Left Books.

Hacking, I. (ed.) (1981) *Scientific Revolutions*, Oxford: Oxford University Press.

Kitcher, P. (1993) *The Advancement of Science: Science without Legend, Objectivity without Illusions*, Oxford: Oxford University Press.

Laudan, L. (1977) *Progress and its Problems*, Berkeley: University of California Press.

Laudan, L. (1984) *Science and Values*, Berkeley: University of California Press.

Newton-Smith, W. (1981) *The Rationality of Science*, London: Routledge.

The sociology of science

Barnes, B., Bloor, D. and Henry, J. (1996) *Scientific Knowledge: A Sociological Analysis*, London: Athlone.

Merton, R.K. (1973) *The Sociology of Science*, Chicago: University of Chicago Press.

Constructivism

Kukla, A. (2000) *Social Constructivism and the Philosophy of Science*, London: Routledge.

Part II

Realism and antirealism about science

5

Scientific realism

However we resolve the debate about the nature of the scientific method, most parties seem to accept that science is usually the best guide we have to the future behaviour of things we can observe, for example comets, bridges, power plants and rainforests. Our scientific knowledge is fallible, partial and approximate, but usually it is the most reliable means we have for predicting phenomena in the world around us. However, science is often taken as telling us much more than this. The natural sciences seem to tell us about the ultimate nature of things and are often thought to have replaced metaphysics as the study of the fundamental structure of reality. Modern science presents us with what seems to be a detailed and unified picture of reality that describes the composition of things and the laws that they obey, from the internal structure of atoms to the life cycles of stars. Contemporary genetics and brain science even seem to offer the prospect of a physical science of human beings and their behaviour. Many of the entities postulated by modern science, such as genes, viruses, atoms, black holes, and most forms of electromagnetic radiation, are unobservable (at least with the unaided senses). So, whatever the scientific method is and however scientific knowledge is justified, we can ask whether we ought to believe what science tells us about reality beyond the appearances of things. Roughly speaking, scientific realism is the view that we should believe in the unobservable objects postulated by our best scientific theories.

Of course, many of those who defend scientific realism also defend the rationality of scientific theory change against sceptics and relativists. However, some ancient and modern critics of scientific realism have not questioned the success or even the progress of scientific inquiry. Many antirealists about scientific knowledge in the history of philosophy are happy to agree with realists that science is the paradigm of rational inquiry, and that it has produced a cumulative growth of empirical knowledge. However, antirealists of various kinds place limits on the extent and nature of scientific knowledge. Hence, the issue of scientific realism is more subtle than many of the polarised debates of science wars, and it is important not to confuse the former with questions about the rationality of science.

The disputes about scientific realism are closely related to those about other kinds of realism in philosophy, some of which will be explained in this chapter, but the reader – especially one with a good deal of scientific knowledge – may already be feeling impatient. Isn't it just obvious that plenty of unobservables described by scientific theories exist; after all, don't scientists manipulate things like atoms and invisible radiation when they design microchips and mobile phone networks? In fact, is it really correct to describe atoms as unobservable? After all, don't we now see photographs of crystal lattices made with microscopes that use electrons instead of light to generate images? Is there really any room for reasonable doubt that atoms exist when so many different parts of science describe how they behave and give rise to everything from the characteristic glow of the gas in a neon light on a billboard, to the way that haemoglobin in red blood cells absorbs oxygen in our lungs?

Even if we decide that atoms are now observable, the issue of principle returns when we ask about the existence of the entities that supposedly make up atoms, and so on. Furthermore, scientists of the past claimed to be manipulating and observing theoretical entities that no longer feature in our best scientific theories, so why should we have such faith that we have it right this time? These and other arguments for and against scientific realism will be the subject of the chapters that follow. First, in this chapter, I will explain the background of the contemporary debate, and the different components of scientific realism. We begin with the distinction between appearance and reality.

5.1   Appearance and reality

The physicist Arthur Eddington makes the distinction between appearance and reality conspicuous with his famous discussion of two tables:

> One of them has been familiar to me from earliest years. It is a commonplace object of that environment which I call the world. How shall I describe it? It has extension; it is comparatively permanent; it is coloured; above all it is substantial.
>
> (Eddington 1928: ix)

> Table No. 2 is my scientific table. It is a more recent acquaintance and I do not feel so familiar with it. It does not belong to the world previously mentioned – that world which spontaneously appears around me when I open my eyes, though how much of it is objective and much subjective I do not here consider. It is part of a world which in more devious ways has forced itself on my attention. My scientific table is mostly emptiness. Sparsely scattered in that emptiness are numerous electric charges rushing about with great speed; but their combined bulk amounts to less than a billionth of the bulk of the table itself.
>
> (Eddington 1928: x)

Eddington distinguishes between the world of common sense and the world as it is described by science. The scientific description suggests that common-sense reality is an illusion, or at least that we certainly don't perceive the world to be anything like the way it is in certain respects. In the twentieth century, physics became increasingly abstract and removed from common sense. In particular, relativity theory and quantum mechanics made the scientific understanding of space and time and the nature of matter, respectively, remote from everyday experience. The description of the ultimate constituents of the table given by contemporary physics depends upon a lot of very difficult mathematics; it is not possible to understand the multi-dimensional worlds of quantum fields, 'superstrings' and the like without it. Hence, although there is an everyday counterpart to the scientific table, there are no everyday counterparts to the 'electric charges' that compose it. So do both tables really exist? If so what is the relationship between them? To understand the philosophical issues raised by Eddington's two tables we must again return to the scientific revolution, and to a philosophical distinction between two types of property that was employed by many of the great thinkers who pioneered the modern scientific outlook, namely that between *primary* and *secondary* properties (or qualities).

As we saw in Chapter 1, the scientific revolution was characterised by various features, including:

(1) A renewed emphasis on experiments and the use of novel technologies such as the telescope, the microscope and the air pump to 'torture nature for her secrets'.

(2) The abandonment of much of the qualitative description of nature given in Aristotelian science (for example, the effects of opium were explained by saying opium has a 'dormative virtue'), in favour of quantitative descriptions of natural properties (for example, the idea of an object having a quantity of matter – its mass – rather than a certain virtue of heaviness).

(3) The abandonment of the search for final causes (teleology) characteristic of Aristotelian science and a concentration on immediate (efficient) material causes.

(4) Science was increasingly regarded not as *a priori* knowledge of necessary truths as in Aristotelian *scientia*, but as empirical (*a posteriori*) inquiry.

The guiding image employed by many writers of the time was the depiction of nature as a giant clockwork machine. The point about a clockwork machine is that the parts all work together in harmony, not because they are co-ordinated by mysterious natural motions or final causes, but because each of them communicates its motion with the part adjacent to it by contact. (A cog here turns a cog there which turns the next one and so on.) People began to envisage the possibility of explaining the behaviour of things in terms of motions of the particles that compose them, rather than in terms of essences and 'occult forces'. Mechanics, in the hands of Galileo, Descartes and Newton in particular, became a mathematically precise science of matter in motion, and what happens as a result of collisions between bits of matter. (All of them adopted a principle of inertia, which states that a body continues in its state of motion unless a force acts to change it, so only changes in motion require an explanation.)

Newton's theory of gravitation was problematic because Newton offered no explanation for how the force of gravity was transmitted between bodies separated in space. It seemed that gravity was an example of the kind of 'action at a distance', which mechanist philosophers were trying to avoid. However, the law of gravitation Newton proposed was at least a precise mathematical one and Newtonian mechanics became incredibly empirically successful. The mysterious nature of gravitational attraction was not enough to stop Newton's theory being adopted by increasing numbers of natural philosophers, and eventually it was accepted by all the scientific establishments. In any case, even Newton hoped that mechanical explanation of gravity would one day be available, and the goal of explaining all manner of natural phenomena in mechanical terms was widely shared. *Materialism*, the idea that there is only one substance, namely matter, and that there is no immaterial soul beyond the body because the human mind is no more than the product of matter in motion, was increasingly popular. Meanwhile, there began a scientific research programme to find out how much of human physiology and behaviour could be explained using physics and chemistry. (An important early success was the understanding of the heart as a pump and the discovery of the circulation of the blood due to William Harvey (1578–1657), who had quantitatively analysed the flow of blood.)

To return to the clock analogy, which Locke used to illustrate the goal of natural philosophy as he saw it: the hands seem to move in a co-ordinated way and the chimes ring out the hours, half hours and so on as appropriate; this corresponds to the appearances of things, the observable properties of, say, a piece of gold. However, the clock has inner workings and this mechanism produces the outer appearance of the clock; similarly, the gold has an inner structure that gives rise to its appearance. The goal of natural philosophy is to understand the inner mechanisms responsible for what we observe.

The success of science since Locke's time seems to depend upon the ingenious invention of devices of all kinds to improve upon the accuracy of the senses, say by using scales to measure mass, and also to measure a variety of properties, such as electrical potential, which are not apparent to the senses at all. Indeed, the growth of many sciences depends upon reducing reliance on the specific character of sense experience as a way of gathering data. This can be seen in the history of chemistry, as the categorisation of substances according to colour, smell, and so on, has been gradually replaced in favour of measures such as refractive index, atomic number and ionisation potentials. This offers to fulfil the aspiration for science to be object-ive, if the subjectivity of human perception can be banished from observation in science, in favour of thermometers, light meters, and eventually even automated recording devices. (Of course modern sci-entists often use computers to process further the data that their machines gather, to speed up and cross-check calculations, and to generate directly numerical outputs, graphs, maps and so on.)

To put it simply, primary properties are those properties that things not only appear to have, but which they also have in reality. Second-ary properties are those which things appear to have but which they don't possess in themselves, only in the mind of the observer. There are many arguments to establish a distinction between the properties things really have and those they only appear to have. There are also many arguments to show that we have no principled way of adjudi-cating rival claims about what properties things possess. Most appeal to the relativity and variability of how things seem to different people at different times. If the shape and colour of the table are different depending on the lighting and the position of the observer, who can say what shape and colour the table really is? (Many of these argu-ments were formulated and compiled by the ancient Greek sceptics.) Berkeley gives a famous example of three pans of water, one very cold, one very hot and one at room temperature. If you put your hand in the hot water and then in the room temperature one, the latter feels cold, whereas if you put your hand in the cold one first and then the room temperature one, it feels warm. Hence, the warmth that you feel does not directly correspond to any property of the water. Simi-larly, it seems that modern science tells us that colour vision is the product of a complex process of light refraction, and that the colours we see do not correspond to simple properties of objects.

The distinction between primary and secondary properties goes back at least to the ancient Greek atomists, who thought that things only *seem* to be, for example, sweet to the taste, cold to the touch, or pleasing on the eye, but that these are not the properties of things as they really are. Atomists go on to say that the properties things really have are the properties of the atoms that make them up, plus complex structural properties due to the arrangement of the atoms. Similarly, in the seventeenth century, many advocates of the new mechanical philosophy of the day, such as Locke, Robert Boyle (1627–1692), Pierre Gassendi (1592–1655), and Newton argued that the primary properties of things are those possessed by the corpuscles or particles that compose an everyday object like a table, whereas the secondary properties are caused by the way the corpuscles are organised, but are not really properties of those corpuscles themselves. So, for example, the corpuscles that make up my table are not brown, nor are the corpuscles making up honey sweet. The colour of a table or the taste of a food are not primary but secondary qualities of those things. On the other hand, the corpuscularians of the seventeenth century argued, corpuscles do all have a shape, a position and are either moving or stationary, so these are some of their primary properties. Among these and other 'mechanical philosophers', there was a general consensus that science should focus on the primary properties of things in order to explain how things appear to us.

Locke distinguished between the *real* and *nominal* essences of things. The nominal essence of gold is just the abstract and general idea that we have of it; so it is yellow, heavy, malleable, dissolves in certain acids, is shiny, and so on. The nominal essence is based on the appearance of gold to us, but of course there are other things that appear like gold, such as iron pyrites or 'fool's gold', and sometimes real gold may not conform to the nominal essence, for example, if it is molten. Hence, what distinguishes real gold from fool's gold is that the former has the real essence of gold and the latter does not. The real essence of something is whatever its underlying nature is. Locke saw no evidence that the scientists of his day could be said to 'know' the real essences of things. However, he did think that there was the realistic prospect of 'probable opinion' about them, and he thought that the real essences of things would turn out to be their micro-structural constitution – in other words, the forms and configurations of corpuscles. Modern science seems to have fulfilled this ambition in some cases. For example, the real essence of gold seems to be that it has atomic nuclei consisting of 79 protons.

135

Does a piece of gold have a colour in the dark? If we chop up the gold into very tiny pieces they remain gold, but they no longer look like gold. On the other hand, Locke thinks the bits of gold retain their mass, their impenetrability and spatial extension whatever we do to them and whether we are looking at them or not. Hence, Locke argues that the colour that we perceive when we look at gold does not resemble anything in the object before us. The gold only has the power to bring about characteristic experiences of seeing gold under certain conditions. He concludes the primary properties exist in objects whether we perceive them or not, but secondary properties do not exist unperceived. So, although in a sense the gold does possess the property of yellowness, this is really a *power* or *disposition* to produce a certain type of sensation in us and there is nothing resembling our experience of yellow in the corpuscles themselves. This makes secondary properties such as colours similar to a property like the fragility of a piece of glass, which is a disposition to break under certain conditions, which glass has in virtue of its micro-structural constitution. Glass is fragile even if it is never broken; similarly the table is brown even if nobody looks at it, because it has a stable disposition to appear as it does to a human observer.

Some of the ideas we have of the properties of objects, for example length or volume, resemble the property in the object that causes the idea. These are primary properties. But the corpuscles making up, say, a piece of gold, are not themselves yellow, malleable, shiny, and smooth. Our sensory experience of these properties is produced by the nature, arrangement and state of motion of the corpuscles that make up gold. The idea of, say, yellow, which we get from perception does not resemble the properties of the corpuscles of gold that cause it. Hence, to be more accurate, primary properties are those that resemble our perceptions of them, whereas secondary properties do not. Note that two objects with all the same primary properties must have all the same secondary properties too. This does not work the other way round, because it is possible for things having very different primary properties to give rise to the same secondary property; after all lots of very different substances all interact with the light that hits them to produce the experience of seeing yellow when the light they reflect hits our eyes. (Secondary properties are said to *supervene* on primary properties because of this kind of one-way dependence.)

Primary quantities are supposed to be measurable and quantifiable properties, such as volume, mass or velocity, or at least to be calculated from other quantities, such as density, which is mass divided by volume. In the seventeenth century, the new quantitative way of describing the world was based on the use of geometry to represent the motion of matter in space. Nearly all of the properties of things that were thought to be primary during the Scientific Revolution, for example extension, motion and size, can be represented geometrically. Calculus enabled Newton to calculate velocity and acceleration geometrically too. Descartes also believed in the distinction between primary and secondary properties. However, he did not believe in atoms, but rather that space is completely full of matter, and so he thought the properties of impenetrability and mass were unnecessary. He also thought all the primary properties were geometrical, but the non-geometrical primary property of mass was eventually widely accepted. Since then, science had relied increasingly upon properties that can be represented numerically, and related by laws and equations that are mathematically complex.

So Eddington's scientific table is the bearer of the primary properties that are measured and described by scientific theories, whereas the commonplace table also bears the secondary properties of our everyday experience. The secondary properties of the commonplace table are reducible to the primary properties of the scientific table in the sense that they are powers or dispositions to produce, in the right conditions, certain effects in us. For example, the brownness of the table is a disposition to produce a characteristic kind of sensory experience in us under normal lighting conditions. Once we adopt the primary/secondary property distinction we need to explain the relationship between our experience of things and their primary properties, and also how we can *know* about the primary properties of things at all. If we concede that many of our ideas of the properties things have do not correspond to their real properties, how do we know that our ideas of the supposed primary properties of things correspond to how those things are? Furthermore, how do we even know there are any things beyond our experiences?

5.2   The metaphysics of the external world

The debate about scientific realism is closely related to the general issue of our knowledge of the external world in the history of philosophy. **Metaphysical realism** is the view that our ordinary language refers to, and sometimes says true things about, a *mind independent* world; Eddington's commonsense table exists and is heavy, brown, solid and so on. Mind-independence means that even if there were suddenly no human beings or other creatures to perceive tables, those that exist would still do so. In other words, the table in my room is here whether or not anyone looks at it. (Philosophers often talk about the 'external world', as opposed to the internal world of the mind.) Scientific realism involves a metaphysical commitment of a similar nature to Eddington's table No. 2; electrons, genes and other unobservables are part of a mind-independent world.

Many people become frustrated with philosophy when they learn that philosophers worry about such issues. Even if it seems legitimate to question the existence of electrons, how can anyone seriously doubt the existence of everyday objects such as tables, trees and other people? However, the real philosophical problem is not to find out if everyday objects exist, or even if we know that they do, but rather to explain *how* we can know that they exist and what their nature is. The problem of metaphysical realism arises because of the difficulty in making knowledge of the external world compatible with seemingly plausible empiricist theories about knowledge and perception. As with the problem of induction, when a philosophical argument that seems sound leads to a conclusion that most people cannot accept (for example that induction is irrational, or that we can't know there are tables in the world), perhaps the challenge for philosophers is not to persuade people of these absurd conclusions, but to identify where the flaw is in the argument.

5.2.1   Realism and ideaism

It seems that our knowledge of the world comes from our senses. It is only because I perceive the table in front of me by seeing and touching it that I know it is there. The simplest view to take about the world and our perception of it is called **direct realism**.

> **Direct realism**: there are external objects that exist independent-
> ly of our minds and which we directly perceive with the senses.

However, it was argued above that many of the properties that the table seems to have are, in fact, artefacts of the way our sense organs and perception works. Many philosophers have taken the arguments we discussed in the section, among others, to show that we do not directly perceive objects in the world around us, but rather the representations of those things produced in the mind. For example, under a red light, and seen from a distance, the table in front of me looks red and rectangular, even though it is in fact brown and square. If what I see is an image of the table in my mind's eye, as it were, this explains how it is that the table can look different to different observers and under different lighting conditions and so on. Another famous argument to show that the direct objects of perception are not external objects is the argument from illusion: take the case of a straight stick which appears bent when standing in a transparent container of water; the stick itself is not bent, yet the stick we see is bent, therefore what we see is not the real stick, but an image or idea of it. These and other arguments that appeal to cases of perceptual error, dreams and hallucinations may be taken to show that the senses do not give us direct knowledge of objects.

A couple of further reasons for abandoning direct realism derive from our scientific understanding of how the senses work. So, for example, in the case of sight, we know that light is either emitted or reflected by the things that we see, that it passes through space, and then impacts upon the eyeball where an image is focused on the retina. The light hitting the retina activates certain cells that send electrical impulses to the visual cortex of the brain. This is only the beginning of the story but already there are two important features to note; first, there is a chain of causes between an object and a person that sees it, and secondly, time elapses between the object emitting the light and a person seeing something. This is already enough to suggest that perception cannot be direct, because we see things as they were recently not as they are at the time we perceive them, and because perception is mediated by the image produced on the retina.

In Chapter 4, I discussed the idea of theory-laden observation. Recent empirical work seems to suggest that what we see is, at least in part, constructed by our brains, rather than being simply an image transmitted from the retina. So, for example, it has been found that if someone is given a pair of spectacles with lenses that turn everything upside down, then at first they will be quite unable to see properly. However, after a while, their brain will adapt and they will start seeing everything the right way up again. Then if the spectacles are later removed, the world will look upside down again until the brain readjusts.

The doctrine that the immediate or direct objects of perception are ideas in the mind, rather than objects in the external world was called **ideaism** by Alan Musgrave (1993) (not to be confused with **idealism**, of which more later).

> **Ideaism:** We do not directly perceive external objects but rather our minds' own ideas or representations of the world.

This was the theory held by the British empiricists, Locke, Berkeley and Hume, who have done so much to influence philosophy of science. According to these thinkers, the mind is not directly aware of objects in the world at all, but rather of what they called 'ideas' and 'impressions': Locke says the mind 'hath no other immediate object but its own ideas' (Locke 1964: Book IV, i, I); Berkeley says 'the objects of human knowledge are either ideas actually imprinted on the senses, or else such as are perceived by attending to the passions and operations of the mind' (Berkely 1975: Part I, I); and Hume says '[a]ll the perceptions of the human mind resolve themselves into impressions and ideas' (Hume 1978: I,i,I). A twentieth century version of ideaism was maintained by Alfred Ayer (1910–1989) who said, 'one can directly experience only what is private to oneself' (Ayer 1940: 136). (These supposed immediate objects of experience are what many twentieth century philosophers used to call 'sense-data'; they are also sometimes called 'the given'.)

All of the British empiricists thought that there were basically two types of mental objects or ideas, namely those that are produced by the senses and the emotions, and those that are copies or faint images of the former. It is best to take Hume's terminology because he distinguishes between 'impressions' and 'ideas', to make this distinction clear; hence, the impression of red is what is forced on the mind by the senses, whereas the idea of red is the image of that impression that one can bring before the mind at will. Similarly, the impression of anger is the feeling that one has when one is angry, whereas the idea of anger is the faint copy of that impression which is before the mind when one thinks about anger.

Ideaism contradicts direct realism, but not the part of the latter that says that there are external objects. Ideaism is a thesis about the nature of perception, not a metaphysical thesis about what exists. Hence, ideaism is compatible with metaphysical realism. If ideaism is right and we perceive only our own impressions and ideas, then what is the relationship between objects in the world and the impressions I have of them? The obvious answer is that my impressions are *caused* by external objects. The primary properties of those objects resemble the impressions they produce in us, but the impressions of colours, tastes and so on are caused by configurations of the primary properties that do not resemble them. Impressions are the direct objects of our experience, and they stand between us and the external world of objects and their genuine properties. Such a view is called representative, indirect or **causal realism** and is to be contrasted with direct realism.

> **Causal realism**: there are external objects that exist independently of our minds and which cause our *indirect* perception of them via the senses.

Causal realism was advocated by Locke who combined it with the distinction between primary and secondary properties and corpuscularianism.

However, once we adopt causal realism instead of direct realism, we have opened up a gap between the world as we perceive it and the world as it is. How then do we know that our perception of the world is at all faithful to the way the world is? Sceptical arguments known since ancient times make the senses seem a dubious source of knowledge. We have seen how the philosophers of the scientific revolution accepted the idea that reality might be very different from how things appear to the senses. The distinction between primary and secondary properties is intended to separate those aspects of our sensory experience of the world that we can trust as guides to the real properties of things, from those we cannot. However, assuming that some properties are secondary, and do not resemble the ideas they cause in us, how do we know there are *any* primary properties resembling our ideas of them? For Locke, it is down to the practice of science to give us probable opinion about such things. In his day, natural philosophers thought it unnecessary to imagine that anything like yellow as we experience it belongs to objects in the world, in order to explain our experience. However, they did find it necessary to impute primary properties of length, breadth, motion and so on to explain our experience. For them, primary properties are the properties of matter, most fundamentally the property of extension (occupying a volume of space).

5.2.2 *Idealism*

Berkeley famously attacked the distinction between primary and secondary properties. The interesting thing about Berkeley's arguments is that, although his conclusions are radically opposed to those of Locke, he bases his arguments on doctrines that Locke seems to accept, such as empiricism about knowledge and meaning, and ideaism. Berkeley was opposed to causal realism, but also to any other form of metaphysical realism, and he denied the existence of matter. This is enough for many to discount Berkeley as someone who contradicts our common sense. However, it is important to be clear that he does not claim that tables, chairs and other 'material objects' don't exist, but that these and other such objects are not mind-independent, and are not made up of corpuscles that possess primary properties.

His first argument is that the doctrine of materialism is meaningless. Here, materialism is not a view about the reducibility of the mental to the physical, but is just the doctrine that matter exists. 'Matter' means something quite specific, namely that which possesses the primary qualities of extension in space, motion, number and so on, and which exists independently of the mind. Berkeley's argument is as follows:

(1) We experience only 'ideas' and not material objects (ideaism).
(2) All our ideas come from experience (concept empiricism).

(3) The words 'material object' cannot stand for any idea and are therefore meaningless (immaterialism).

Like many empiricists after him, Locke argued that the mind at birth is like a blank sheet of paper which is then written on by experience, and was strongly opposed to the rationalist thesis that we have ideas, or concepts prior to sensory experience of the world. Hence, Locke seems to believe both (1) and (2). But if we cannot experience matter directly, and if all our ideas are derived from experience, then we ought not to be able to have an idea of matter. Matter is defined as that which lies beyond all our experience, so how can experience give us any idea of it? The reader may be tempted to abandon ideaism at this juncture, but as we have seen, the arguments for ideaism are quite strong. However, perhaps a more promising response is to qualify (2). After all, our ideas may all come from experience, but it may still be possible for us to combine them to create the idea of matter, say by combining the idea of extension in space with the idea of a cause. Berkeley has much more to say, for example, about why matter cannot cause anything, but we will leave this argument for now and turn to his attack on the primary/secondary property distinction.

Berkeley denied the distinction between primary and secondary properties on various grounds, which are summarised below:

(a) The primary/secondary distinction is supposed to correspond to that between the *objective* and the *subjective* but, Berkeley argues, nobody has adequately characterised the latter distinction, and so it cannot be invoked to explain the former one.

(b) The primary properties are supposed to be those that are stable, while the secondary properties are those that are perceptually relative; but we cannot know that the primary properties are really stable, only that they are stable relative to our perceptual make-up. Indeed, the size of things is dependent on some scale being adopted – motion, as Galileo discovered, is relative to some frame of reference and so on. Hence, Berkeley argues, primary properties are as relative and variable as secondary ones.

(c) The primary properties of a body are those which it is supposed to have just in virtue of it being a material object, but then our experience of bodies is that they always have a colour as well as a shape and so on. Berkeley denies that one can imagine a material body that has no colour at all, hence he argues that there are no grounds for detaching colours and so on from material objects but not shape, size and motion as well.

What can we say in response to these arguments? We could try and mount a defence of extension, motion and so on as primary properties, but unfortunately, about these properties modern science seems to be on Berkeley's side. None of the primary properties of matter listed by Locke and other corpsucularians are now regarded as true properties of the ultimate constituents of matter. Even mass is now regarded as a secondary property produced by the 'rest mass' of things in a certain frame of reference. The only candidates for primary properties that physical science now ascribes to things, such as charge, isospin, spin, 'colour-charge', and so on, lack any counterparts in our experience, so we can hardly say the sensations they produce in us resemble them. However, we can argue that these are properties of things that they possess independently of our perception, and to that extent rescue the idea of primary properties. If we give up on the idea that the primary qualities of things include extension, shape, motion and so on, and instead accept whatever properties contemporary science includes, then we can defend the intelligibility of the idea that physical objects such as electrons and atoms exist mind-independently against arguments (b) and (c). In response to argument (a) we can argue that the properties revealed by modern science are objective, because they can be measured precisely in a way which is repeatable and independent of who is doing the experiment.

For Berkeley, all the supposed primary properties of matter, including extension, are secondary; in other words, all properties only exist when perceived. (For Aristotle all properties are primary.) Berkeley's positive view was a version of idealism. Idealism is the metaphysical thesis that all that exists is mental or spiritual in nature, hence it is incompatible with any form of metaphysical realism, whether direct or causal. Berkeley took his attack on the idea of matter and on the idea of primary properties to show that nothing other than minds can exist unperceived. He argued that so-called external, mind-independent objects are, in fact, mind-dependent, as follows:

(a) We perceive such things as trees and stones.

(b) We perceive only ideas and aggregates (or collections) of them (Ideaism).

(c) Ideas and aggregates of them cannot exist unperceived.

(d) Therefore, trees and stones are ideas and impressions or aggregates of them, and cannot exist unperceived (Idealism).

This argument is clearly valid. If we only perceive collections of ideas, and we perceive everyday objects, then everyday objects must be collections of ideas. If we want to deny (d) and maintain some form of metaphysical realism, we must deny one of the premises. (c) seems to be right. When I am thinking about a horse, surely my idea of it does not exist independently of my mind. As we have seen, Locke accepts ideaism, (b), but gives up on (a) above, and adopts causal realism. Berkeley thinks that his idealism is more faithful to common sense than Locke's causal realism because Locke denies (a). Berkeley calls this the 'royal road to scepticism', because he thinks that once (a) is abandoned there is no way to be sure that we have any knowledge of objects at all.

Empiricists such as Locke adopt ideaism to secure the infallibility of the 'empirical basis' for knowledge in the face of sceptical arguments that seem to undermine it. However, once we look at things this way it is natural to ask how we know that there are *any* external objects causing our ideas of them. This is the problem of scepticism about the external world that was originally made vivid by Descartes. He imagined there was an evil demon feeding sensory impressions to him that created the illusion of a world of objects and people other than himself, when it fact there were none. These days we are more likely to motivate this very strong form of scepticism using the image of a brain in a vat, with its sensory nerves being stimulated by a mad scientist or a computer program. The question in each case is how can we know that this is not our actual situation? If you can't rule out the possibility that you are a brain in a vat, then how can you know any of the things you take yourself to know about the world around you? If we adopt ideaism it seems that we are stuck behind the 'veil of ideas' and in what Ayer called the 'egocentric predicament'. Like Descartes, philosophers such as Berkeley, Hume, Russell and Carnap have all attempted to start with sense-data and recover the everyday world.

The obvious problem for Berkeley's view is to explain how it is that there is such coherence in our ideas, in other words, why we all more or less agree so much of the time about the properties of things in the world around us, and why objects reappear where they were before when we re-enter an empty room. Berkeley's answer to this depends upon his belief in God. He argues that God perceives everything all the time and hence ensures the continued existence of the world around us when we are not observing it. This is unacceptable to the atheist of course, and anyway seems ad hoc. However, although we may reject Berkeley's conclusions, we must appreciate the strength of his arguments and the difficulties they raise for realism.

The subtler idealism of Kant is a more palatable strategy aimed at avoiding scepticism. Whereas Berkeley collapses the external world onto our impressions to solve the problem of how we get from knowledge of the latter to knowledge of the former, Kant agrees with metaphysical realists that there is a mind-independent world, but he agrees with sceptics that we can't have knowledge of it. Instead, he argues, all our knowledge is of the world as it is *for us*. The world in itself he calls the *noumenal* world, and the world as we experience it he calls the *phenomenal* world. Much of our knowledge is of particular facts about the phenomenal world learned through the senses, but Kant thought that some of our knowledge is *a priori*. Although we can only know the measurements of a specific triangle through our senses, we can know that any triangle we experience will have internal angles summing to 180°, by the use of reason alone. According to Kant, arithmetic, geometry and Newtonian mechanics are *a priori* forms of knowledge, not about the noumenal world, but about the form our experience must take. Unfortunately, as I mentioned in Chapter 2, this unravelled in the face of scientific developments. The new physics of the late nineteenth and early twentieth centuries – in particular, relativity theory and quantum mechanics – seems to refute some of the principles of classical science and mathematics.

5.3 Semantics

The philosophy that seems appropriate for the scientific age is empiricism. Our knowledge of the world does indeed seem to be based not on *a priori* reflection, but on centuries of empirical investigation. However, empiricism is beset by a fundamental problem, namely, if all our contact with the world is mediated by our 'ideas', then how can we know that our experience is a reliable guide to the world as it is in itself. This problem was clearly appreciated by Hume who argued that, although we have no choice but to continue to act as if there is an external world, we have no rational grounds for believing it. Most philosophers have been at least as dissatisfied with Hume's capitulation to scepticism about the external world as with his capitulation to scepticism about induction. We have seen how Berkeley dealt with the problem by denying that there is anything to objects beyond our ideas. According to his view, tables and trees are collections of impressions which have no existence independent of perception. Naturally, such a challenge to common sense is usually met with incredulity by scientifically minded philosophers.

Commonsense metaphysical realism (about the existence of the world around us) is not contentious. Many contemporary western philosophers would say that we are justified in believing in the existence of mind-independent objects, because they are the best explanation of the regularities in our experience. Scientific realists argue that these are just the grounds we have for believing in the existence of the unobservable entities postulated by our best modern scientific theories. Antirealists about scientific knowledge are usually empiricists who oppose the way realists think science can go beyond experience and get at the real causes of things. As we shall see, logical positivism was, in part, an attempt in the hands of some philosophers to put an end to debates about metaphysical realism, antirealism and idealism.

5.3.1 Logical positivism

The term 'positivism' was coined by a French philosopher called Auguste Comte (1798–1857) who argued that societies pass through three stages – namely the theological, the metaphysical and the scientific. In the theological stage, people explain phenomena such as thunder, drought and disease by invoking the actions of gods, spirits and magic. In the metaphysical stage, they resort to unobservable forces, particles and so on. The scientific stage is achieved when pretensions to explain why things happen, or to know the nature of things in themselves, are renounced; the proper goal of science is simply the prediction of phenomena. He aimed to complete the transition of European thought to the scientific stage by advancing the scientific study of society and social relations (sociology), and established a system of rituals celebrating scientists and science, to replace the traditional calendar of Saint's Days and religious festivals.

Positivism has its roots in empiricism, especially in Hume's attempt to separate the meaningful from meaningless (see Chapter 2, section 2.1). In general, positivists:

(a) emphasise verification/falsification;
(b) regard observation/experience as the only source of knowledge (empiricism);
(c) are anti-causation;
(d) are anti-theoretical entities;
(e) downplay explanation;
(f) are, in general, anti-metaphysics.

There is something positivist in spirit about the mechanists of the scientific revolution with their desire to create a science rooted in experiment, which could avoid the mysterious 'essences' and 'virtues' of Aristotelian natural philosophy. Of course, they were also anti-positivist in so far as they posited theoretical entities such as atoms and forces, and invoked them as causes in their explanations. Hume was a positivist in that he was sceptical about: any relation of causation beyond our association of ideas (see 2.1); substance (or matter beyond phenomena); and the soul (or any idea of the self beyond a fleeting flux of ideas and impressions). Later, the physicist Ernst Mach (1838–1916) argued that physical science should only concern itself with what was observable, and that the function of laws in science is simply to systematise the relations between our experiences.

In the nineteenth century, metaphysics thrived. Idealism and romanticism were prevalent in philosophy, and such mysterious notions as 'the absolute', 'becoming', and 'the will' were widely discussed. By the beginning of the twentieth century, a reaction against this kind of philosophy was developing, and many philosophers and others embraced science, mathematics and logic as the antidote to what they saw as the confused thinking of metaphysicians. Logical positivism was originally centred around a group of scientists, mathematicians and philosophers called the Vienna Circle, which met in the 1920s. Many of the Vienna Circle were Jewish and/or socialists. The rise of fascism in Nazi Germany led to their dispersal to America and elsewhere, where the ideas and personalities of logical positivism had a great influence on the development of both science and philosophy.

The difference between logical positivism and logical empiricism is a matter of scholarly dispute. The most influential of those classified as logical positivists or empiricists include Moritz Schlick (1882–1936), Carl Hempel (1905–1997), Carnap, Reichenbach (although he was in Berlin, not Vienna), and Ayer (he visited the Circle and brought some of its ideas to Britain). They all adopted the empiricism of Hume and Mach and Comte's aspiration for a fully scientific intellectual culture. What was new about them was that they exploited the mathematical logic, recently developed by Gottlob Frege (1848–1925) and Russell among others, to provide a framework within which theories could be precisely formulated. The idea was that if the connections between ideas and associated experiences could be made precise, then it would be possible to separate meaningless metaphysical mumbo-jumbo from empirical science.

Most of the words we use are for things that are manifest to the unaided senses. As children we learn the names for objects, properties and processes that appear in the world around us by hearing adults use those names and copying them. Of course, language acquisition is amazing because children seem to be so good at it, and because they master grammar and syntax with the minimum of examples. In principle, however, we can understand how they learn the meanings of words like cat, house, red, square, cooking and running. We can even imagine how they go on to build up a full vocabulary from these simple beginnings, because, once someone has enough language, the meanings of new words can be explained to them in terms of words they already know. I may have never seen a woolly mammoth but I know it is a big, hairy elephant, and so on. Hence, it is natural to think that all words get their meanings by their connection to what we can experience, even though sometimes the connection may be remote (concept empiricism).

On this view, the content of each of our thoughts must somehow be tied to ideas the mind acquires through sensory experience of the world. This implies that no matter of fact that can be intelligibly or meaningfully thought about can go beyond all possible experience. Russell and Ludwig Wittgenstein (1889–1951) are supposed to have had an argument about whether it is meaningful to posit the existence of a hippopotamus that cannot be detected by any scent, sight, sound or touch. If we add to the senses the use of all scientific detection equipment such as radar, infrared cameras and sonar, and the creature is still supposed to be undetectable, then surely it is meaningless to talk about it. Hence, we arrive at what has been called the 'empiricist criterion of meaning', which takes various forms, but is roughly the idea that, to be meaningful, a word must have some connection with what can be experienced.

The logical positivists used this criterion of meaning to criticise theories that they thought were pseudo-scientific, such as psychoanalysis and the theory of vitalism (which posited vital forces responsible for the life of cells and other organisms) because they employed theoretical terms and concepts that could not be explicitly related to what can be observed. They also criticised metaphysics for being meaningless, hence Carnap said:

> The metaphysician tells us that empirical truth-conditions [for metaphysical terms such as 'the absolute'] cannot be specified; if he asserts that nonetheless he 'means' something, we show that this is merely an allusion to associated words and feelings, which however, do not bestow a meaning.
>
> (Carnap 1959: 65)

Similarly, many positivists argued that theological hypotheses were meaningless. For example, since the hypothesis that God is perfectly good or omnipotent does not imply anything about what we experience with our senses, it is, strictly speaking, meaningless. A selection of what they called 'psuedo-concepts' includes 'essence', 'thing in itself', 'the Good' and 'the absolute'. They are psuedo-concepts because statements containing them assert nothing. On the other hand, take the term 'radio wave'. Positivists argue that what makes this term different to a term like 'the absolute' is that its use has implications for what we can observe. So for example, the sentence 'there are radio waves passing through this room' implies that I will get a response from a radio receiver by tuning it to the appropriate wavelength.

In Chapter 2, I explained Hume's distinction between relations of ideas and matters of fact. Recall that statements that concern only the relations among ideas are those that are true or false just because of the meanings of words. For example, 'if Jimmy is taller than James, and James is taller than Dawn, then Jimmy is taller than Dawn' is true because of what 'taller than' means. On the other hand, 'Jimmy is taller than James' concerns a matter of fact, because it is true or false depending on the heights of the actual people referred to by the names 'James' and 'Jimmy'. The logical positivists employed Kant's similar distinction between analytic and synthetic statements. Analytic statements include: 'what will be will be', 'trees are plants', and 'red is a colour'. Examples of synthetic statements include: 'Paris is the capital city of France', 'the poles of the Earth are covered with ice', and 'the table is brown'.

Logical positivists had the following basic commitments:

(1) Science is the only intellectually respectable form of inquiry.
(2) All truths are either: (a) analytic, *a priori* and necessary, in other words, tautological, or (b) synthetic, *a posteriori* and contingent.
(3) So far as knowledge goes, it is either purely formal and analytic, such as mathematics and logic, or it is a kind of empirical science.
(4) The purpose of philosophy is to explicate the structure or logic of science. Philosophy is really the epistemology of science and analysing concepts.
(5) Logic is to be used to express precisely the relationships between concepts.
(6) The *verifiability criterion of meaning*: a statement is literally meaningful if and only if it is either analytic or empirically verifiable.
(7) *The Verification Principle*: the meaning of a non-tautological statement is its method of verification; that is, the way in which it can be shown to be true by experience.

The positivists tried to find certain foundations for our knowledge of the world. Consider these criteria for foundational truths:

- They must not be inferred from any other beliefs, but rather be self-evident or self-justifying.
- They must be immune from scepticism.
- They must be useful and informative, in other words synthetic, not analytic.

The idea of **foundationalism** is that the justification of beliefs is of two kinds: some (basic) beliefs are justified independently of any other beliefs, whereas non-basic beliefs are justified because basic beliefs either deductively or inductively entail them. Many empiricists take our knowledge of our sensory states to be foundational. For example, my belief that it seems to me as if the light is on is self-justified. The logical positivists tried to use 'protocol statements' as the foundation of knowledge. These are statements that refer only to the immediate content of some experience or observation, such as 'I see a red light flash'. Protocol statements are also called sense-data reports, or basic propositions (Ayer). They are supposed to be *first person singular, present tense, introspection reports*. That is, reports of how things seem to an observer at a given time. As such they are supposed to be: synthetic and contingent, since the observer's experience could be different; immune from doubt, because anyone is supposed to be able faithfully to report at least how things *seem* to them; and not inferred from other beliefs, because the observer simply reports the experience. Hence, they are supposed to meet the above criteria for foundational truths.

The logical positivists argued that all meaningful empirical statements are either *protocol statements* or *empirical hypotheses*. Empirical hypotheses relate protocol statements to each other, and hence allow for prediction. Scientific laws are empirical hypotheses, and they are tested by the predictions they make about what will be observed. Protocol statements are *strongly verified* because their truth is established conclusively by experience. The problem of induction means that any number of experiences consistent with the predictions of a particular law or generalisation, such as all metals expand when heated, do not make it certain that the next observation will follow the same pattern. So protocol statements at best only *weakly verify* empirical hypotheses in the sense that they make them probable rather than certain.

Suppose that we have certain knowledge of the contents of our immediate sensations, and suppose also that empirical hypotheses predicting relations between phenomena are confirmed by observations. This much is consistent with scepticism about the mind-independent world, or with Berkeley's idealism. Recall from 2.2 above that the problem is: granting that we can know about regularities in our experience, how can we know there are objects and people in the world causing them. In other words, even if we can know empirical hypotheses, as well as analytic truths and protocol statements, how can we build knowledge up from these? In other words, how can collections of private sense-data conceivably make up a common world?

We seem to be faced with a dilemma:

(a) We know lots of things about cabbages and kings.
(b) We only know protocol statements and analytic truths.

Scepticism denies (a) – the logical positivists wanted to infer (a) from (b). However, because protocol statements are only certain because they do not refer to anything beyond immediate experience, we cannot use a deductive argument to get from protocol statements to a statement about mind-independent objects, since the conclusion must be somehow implicit in the premises for deductive arguments to be valid. On the other hand, we cannot use an inductive argument to infer (a) from (b), since to do so we would need to observe the coincidence of sensations with the existence of objects, which requires independent access to such objects.

The positivists adopted a lateral thinking solution; they reduced all knowledge to knowledge of protocol statements and necessary truths. They argued that talk of perceived or possible objects is *reducible* to talk of actual or possible experiences. Propositions asserting the existence of physical objects are equivalent to ones asserting that observers will have certain sequences of sensations in certain circumstances. Physical objects are logical constructions out of actual and possible sense-experiences. A physical object is 'a permanent possibility of sensation' and nothing more. This view is sometimes called **phenomenalism**: the logical positivists agreed with those who think philosophers debating the existence of the table is ridiculous; talk of the external world is literally meaningless.

So it seems foundationalism and the radical empiricist doctrine of ideaism together suggest that, if we want to avoid metaphysics, phenomenalism is the only way out of scepticism. So much for Eddington's everyday table; it is a mere construct out of sense-data. But what about the scientific table and all its atoms, electrons, forces and so on? The basic positive goals of logical positivism were as follows:

(I)   To show that the use of theoretical terms in science is consistent with the empiricist criterion of meaning.
(II) To show how statements about observations can confirm theoretical statements, in other words to explicate the 'logic of confirmation'.
(III) To show that mathematics and logic are analytic.

We will return to confirmation in later chapters. Although the history of the attempt to achieve (III), and the philosophical issues it raised, are among the most important parts of twentieth century philosophy, they will not concern us further (see further reading below). In the next section we will be concerned with (I).

5.3.2   Semantic instrumentalism and reductive empiricism

For positivists, all empirical knowledge aims at the successful anticipation of future experience. Nineteenth century positivists such as Comte and Mach thought that the postulation of unobservable atoms and fields had no place in science. The problem the logical positivists faced was that the science they admired so much seemed to make indispensable use of empirical hypotheses and laws involving theoretical terms such as 'atom', 'charge', 'nuclear force' and so on. If concept empiricism is right, then how can such terms be meaningful? The logical positivists had two approaches to what they considered to be empirically dubious; they either denied that it was meaningful, as with metaphysics, ethics, theology and so on, or they showed how the meaning could be empirically sanitised to eliminate metaphysics, as with our knowledge of the external world. In order to explain why people talk about supposedly meaningless things so much, Ayer argued that saying 'murder is wrong' merely expresses the speaker's attitude to murder, rather than stating anything that could be true or false. This is sometimes known as the 'boo–hurrah' theory of ethics, according to which all statements about ethics are merely expressions of an emotional reaction with no other content. Philosophers use the term *assertoric* to mean 'genuinely asserts something about the world'. Only assertoric statements are capable of being true or false; for example, the statement 'welcome' is not assertoric, but it does express an attitude. For a statement to be assertoric it does not have to be true; when I say Bristol is bigger than London, I genuinely assert something, but it is false.

Similarly, some antirealists about science deny that statements involving theoretical terms are assertoric. Just as we can use the concept of the average tax-payer to simplify thinking about economics, without thinking that anyone who is the average tax-payer exists, so we can talk about electrons and so on without taking it all literally as really referring to something. This is called **semantic instrumentalism**.

> **Semantic instrumentalism**: the theoretical terms of scientific theories should not be taken literally as referring to unobservable entities, because they are merely logical constructs used as tools for systematising relations between phenomena. Theoretical hypotheses are not assertoric.

On the other hand, others argue that statements involving theoretical terms are assertoric, but that what they say is reducible to statements about observables. The conjunction of (6) and (7) of 5.3.1 above implies that if the theoretical terms of modern science are really meaningful, it ought to be possible to define each of them using only words that refer to sensory experiences and everyday observable objects.

> **Reductive empiricism**: theoretical terms can be defined in terms of observational concepts, hence statements involving them are assertoric. Scientific theories should not be taken literally as referring to unobservable objects.

Here we have an explicit definition of a theoretical term:

$$\text{Pressure} = \frac{\text{Force}}{\text{Area}}$$

The problem is most scientific terms cannot be defined this way. Take the term 'temperature'. We need to give an *operational definition* for each possible temperature, since we are not allowed to think of there being one real property with different possible values. An operational definition of 'the temperature is 100° Celsius' is that it is applicable when water is observed to boil at normal atmospheric pressure; but it is also true that a properly calibrated mercury thermometer will read 100, and so too will an alcohol thermometer. The good thing about using a single property of 100° Celsius is that it systematises relations between different experiences. However, according to the operationalist understanding of theoretical terms, every time we encounter a new way of associating the attribution of a temperature of 100° Celsius to something, with something concrete in our experience, we are really introducing a new property. This makes it nonsense to talk about theories being extended to deal with new phenomena. Yet the history of science is full of cases where a theory that was introduced to account for one domain enjoyed empirical success in another one. For example, Maxwell's theory of the electromagnetic field was introduced to account for electricity and magnetism, but it unexpectedly predicted the existence of electromagnetic waves, and thereby accounted in part for the behaviour of light, and also X-rays, infrared, microwaves and so on. (Carnap saw this problem and gave up on seeking an explicit definition of theoretical terms. However, he still wanted to anchor them to experience by giving them a partial interpretation in terms of so-called correspondence rules (see Psillos 1999: chapters 1–3).)

The positivists soon realised that once we describe our experiences in a way that uses a public language then sense-data reports are not incontrovertible, because we have to apply terms correctly and mistakes are possible. In the case of the real observational language of science, say 'the reading on the ammeter was 4 amps', it is obvious that observation reports are fallible. That is why scientists repeat tests that other people have done to guard against error. The logical positivists and empiricists reacted to this problem by giving up on certain foundations of knowledge and abandoning the verification principle ((7) of 5.3.1 above). They emphasised instead empirical *confirmation* or *falsification*, in other words, something counting for or against the proposition (empirical decidability). Some of them embraced realism of some kind. An important part of scientific realism is a commitment to taking the language of science, even the parts of it that seem to refer to unobservable entities, literally, rather than trying to reduce or eliminate theoretical terms in favour of observational terms.

5.3.3   Truth

Some forms of antirealism are based, not on the elimination of theoretical terms, but on theories of truth that deny the realist conception of truth as correspondence between language and the world. Hence, some antirealists argue that theoretical statements of science are to be taken literally and are assertoric, but that what makes them true or false is not an objective reality beyond all experience. A social constructivist (see 4.6), for example, need not deny that theoretical terms 'refer' nor that theories are true, but may insist that truth is *internal* to our norms and practices, and that the entities to which we refer are socially constructed. For example, it is perfectly true to say that checkmate is the end of a game of chess, or that England borders Scotland, but these facts are constituted by our conventions. Some philosophers think many or all facts are like that, and apply this to other domains rather than just to science. Hence, a social constructivist about mathematics will argue that there are no truths in mathematics beyond those that are provable by mathematicians.

Hence, we can distinguish various theories of truth, the first being:

The correspondence theory of truth: a statement is true when it corresponds to the facts. The terms in the statement refer to things and properties in the world. The conditions under which statements are true or false (truth-conditions) are objective, and determine the truth or falsity of those statements depending on how things stand in the world.

Most advocates of the correspondence theory regard it as necessary to make explicit a commitment to mind-independent truth conditions. Those who adopt the verification principle ((7) of 5.3.1) believe the idea of truth beyond what can be verified makes no sense. Others argue for a *pragmatic* theory of truth, which is one according to which what is true is something like 'what works best in the long run'. Others that, because we can never escape our language and tell whether it successfully 'hooks onto the world', what is true is what fits best with everything else we believe; this is a *coherence* theory of truth.

5.4 Standard scientific realism

If we incorporate both metaphysical and semantic realism about some subject matter S (which could be ethics, mathematics, aesthetics or theoretical science, among others) and add an *epistemic* requirement we get a strong form of realism about S:

(i)   the entities or kinds of entities talked about and/or described by discourse about S exist;
(ii) their existence is independent of our knowledge and minds.

These are the metaphysical requirements.

(iii) Statements about S are irreducible/ineliminable and are genuinely *assertoric* expressions;
(iv) truth conditions for statements of S are objective and determine the truth or falsity of those statements depending on how things stand in the world.

These semantic requirements are cashed out in terms of a correspondence theory of truth, as opposed to a pragmatic or a coherence theory of truth.

(v)   Truths about S are knowable and we do in fact know some of them, and hence the terms of S successfully refer to things in the world.

This is the epistemic requirement.

If we take S to be science we get the classic statement of scientific realism, due in its first modern incarnation to Hilary Putnam (1926–), Wilfrid Sellars (1912–1989) and others.

For example, if we are considering electron theory then scientific realism says that:

(i)   electrons exist;

(ii) mind-independently;

(iii) statements about electrons are really about subatomic entities with negative charge, spin 1/2, a certain mass, and so on;

(iv) these statements are true or false depending on how the world is;

(v) we should believe electron theory and much of it counts as knowledge.

5.5 Antirealism

We have seen that scientific realism involves three kinds of philosophical commitment: a metaphysical commitment to the existence of a mind-independent world of observable and unobservable objects; a semantic commitment to the literal interpretation of scientific theories and a correspondence theory of truth; and finally an epistemological commitment to the claim that we can know that our best current theories are approximately true, and that they successfully refer to (most of) the unobservable entities they postulate, which do indeed exist. To be an antirealist about science it is only necessary to reject one of these commitments, and antirealists may have very different motives, so there are a variety of antirealist positions that we ought now to be able to distinguish: sceptics deny (i), reductive empiricists deny (iii), social constructivists such as Kuhn (on some readings) deny (ii), while, as we shall see in the next chapter, constructive empiricists such as Bas van Fraassen deny only (v), but also do not believe or remain agnostic about (i).

—◦◯◦—

Alice: So now you're saying that science might give us knowledge up to a point but it only tells us about what we can observe?

Thomas: Maybe so. It seems possible.

Alice: Yes, well, it's possible that the table we are sitting at is a figment of our imaginations or that it disappears when nobody is looking at it but so what? You can't prove anything beyond doubt but that doesn't mean we don't know anything. If all you are saying is that I have as much right to believe atoms are real as I do to believe the table is real then I agree with you.

Thomas: Slow down. There's a difference. When you claim to know there's a table there, you aren't claiming to know about ultimate reality or the hidden nature of things, just about how things seem.

Alice: Well, I am claiming that the table exists even when I am not looking at it and that it is the same table you see, and that it will still be here if we go away for a minute and then come back and . . .

Thomas: Yes, but at least sometimes we can observe the table. The point about atoms and the like is that they are purely theoretical. For all we know there could be quite different things causing what we see.

Alice: You might as well say that it just looks as if I am sitting here but I'm not really.

Thomas: I don't think it's the same thing, and anyway, as far as science is concerned, all that matters when it comes down to it is getting the predictions right for what we observe. Lots of different theories that disagree about what the unobservable world is like could still agree in what they predict about the results of experiments.

—o☉o—

Further reading

Realism, ideaism and idealism

Berkeley, G. (1975a) 'The principles of human knowledge', in M.R. Ayres (ed.) *Berkeley Philosophical Works*, London: Everyman.

Berkeley, G. (1975b) 'Three dialogues between Hylas and Philonus', in M.R. Ayres (ed.) *Berkeley Philosophical Works*, London: Everyman.

Locke, J. (1964) *An Essay Concerning Human Understanding*, Glasgow: Collins.

Musgrave, A. (1993) *Common Sense, Science and Scepticism: A Historical Introduction to the Theory of Knowledge*, Cambridge: Cambridge University Press.

Woolhouse, R.S. (1988) *The Empiricists*, Oxford: Oxford University Press.

Logical positivism

Ayer, A.J. (1952) *Language, Truth and Logic*, Cambridge: Cambridge University Press.

Friedman, M. (1999) *Logical Positivism Reconsidered*, Cambridge: Cambridge University Press.

Hanfling, O. (ed.) (1981) *Essential Readings in Logical Positivism*, Oxford: Blackwell.

Logicism about mathematics

Shapiro, S. (2000) *Thinking About Mathematics*, Chapter 5, Oxford: Oxford University Press.

6

Underdetermination

What the antirealist needs is a good argument. Theoretical reductions of everyday objects to sense-data lack any interest unless we understand the arguments that lead to them. However, it seems that instrumentalists and reductionists about scientific knowledge have been overwhelmed by history. Contemporary science makes essential use of a host of theoretical terms, and many scientists seem to think that they are really manipulating genes, molecules and electrons. Why should anyone take the antirealist about science seriously? As we have seen, there are many ways to be an antirealist about science if you are some form of antirealist about everything. However, we are interested in the arguments specific to science. In particular, once again it is the fact that science is supposed to tell us about a reality beyond the appearances that will concern us. In the second part of this chapter I will explain constructive empiricism, which is a form of antirealism that denies the epistemic component of scientific realism. First, I will discuss an argument that has long been an important motivation for scepticism about scientific knowledge.

6.1 Underdetermination

All underdetermination arguments exploit the fact that often more than one theory, explanation or law is compatible with the evidence. Data underdetermine the correct theory when the data are insufficient to determine which of several theories is true. This happens all the time in everyday life and in science. Why is the train late? It could be that there is a problem with the engine, a staff shortage, a signal failure and so on. We often suspend judgement in such cases, but sometimes we have to make decisions in the face of uncertainty. For example, doctors have to take some action, even though the cause of a serious condition is often underdetermined by the symptoms; a terrible pain in the stomach could be appendicitis, or it could be a quite different infection. Of course, experience and further investigation enable people to refine their judgements, but everyone will agree that there are some cases where the best the available evidence can do is narrow down a gamble between two or more possible explanations.

Similarly, in science, sometimes several hypotheses predict and explain some phenomenon, and all the observations that have been made are consistent with all of them. As I mentioned in 4.4, this was the case with Copernicus' theory of planetary motions and the Ptolemaic alternative for a while. Given the accuracy of the observations at the time, each theory entailed that the planets and the Moon would appear in the same region of the sky. Yet the theories fundamentally disagree about the real situation; one says that the Earth is at the centre of the solar system, and the other that the Sun is. If, as the semantic component of scientific realism demands, we are to take the theories literally, rather than treating their theoretical terms as mere instruments, then how can we know which one to believe in such circumstances? In the philosophy of science, many have argued that we should suspend judgement as to the real causes of things because of this problem.

6.1.1 *Weak underdetermination*

The idea is to argue as follows.

(1) Some theory, T, is supposed to be known, and all the evidence is consistent with T.
(2) There is another theory T# that is also consistent with all the available evidence for T. (T and T# are *weakly empirically equivalent* in the sense that they are both compatible with the evidence we have gathered so far.)

(3) If all the available evidence for T is consistent with some other hypothesis T#, then there is no reason to believe T to be true and not T#.

Therefore, there is no reason to believe T to be true and not T#.

Pairs of theories that have been alleged to be empirically equivalent include: Ptolemaic and Copernican astronomy between 1540 and early in the seventeenth century; Newtonian and Cartesian physics before the mid eighteenth century; wave and particle optics in the eighteenth century; and atomism and anti-atomism between 1815 and 1880. There are also many examples of weak empirical equivalence in modern science. Sometimes scientists conclude that the two theories are really different version of the same underlying theory (this happened with the two early versions of quantum mechanics), but sometimes the theories really are incompatible.

A variant of underdetermination is the 'curve fitting problem' (see Figure 4). Suppose scientists are interested in the relationship between the pressure and the temperature of a gas at a fixed volume. An experiment is performed and the data points are plotted on a

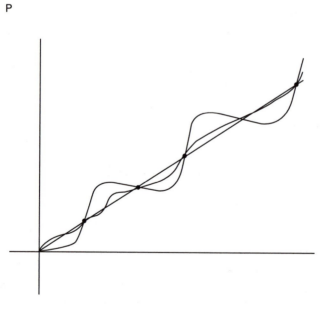

Figure 4

graph. Perhaps it seems obvious that we have a linear relationship between two physical quantities, such that increasing the pressure increases the temperature and vice versa, as shown by the straight line. However, the points we have so far are compatible with the other curves as well. Indeed, through any finite set of data points there are an infinite number of curves on which all of them will lie.

Consequently, one form of the underdetermination argument says that as all the data we have gathered to date are consistent with more than one theory, we ought to suspend judgement as to which theory is true. Call this the *weak form of the underdetermination argument*. This kind of underdetermination problem is faced by scientists every day, and what they usually try and do to address it is to find some phenomenon about which the theories give different predictions, so that some new experimental test can be performed to choose between them. The gathering of more data, or more precise data, may solve weak underdetermination. For example, in the case of astronomy, telescopes were improved so that differences in the predictions of the Ptolemaic and Copernican theories for the motions and appearances of the planets, and their moons, became observable (see Chapter 4). (In the case of the curves above, we simply make more observations to see if they give us points near or on the straight line.)

However, the argument above can be made more powerful so that whatever theory we end up with after we have gathered more data will itself be weakly empirically equivalent to another theory or theories (recall with the curve fitting problem, any set of data points lie on an infinite number of curves).

For any theory H there is *always* another theory G such that:

(i)   H & G are weakly empirically equivalent.
(ii) If H & G are weakly empirically equivalent then there is no reason to believe H and not G.

Therefore, there is no reason to believe H and not G.

This is potentially a real problem for the scientific realist because, if it is correct, there are always rival theories we have not thought of, which fit all the data that support each of the best current scientific theories. If this is the case, why should we believe our best theories and not the sceptic's alternatives?

However, this argument may be challenged by denying the second premise, in other words, by arguing that the mere existence of a rival hypothesis consistent with all the data so far does not mean there is no reason to prefer one of H and G. Hence, for example, Popper argued that if G is ad hoc, and entails no other empirically falsifiable predictions, then it should be ignored (see 3.5, pp. 88–89). Of course Popper didn't think we should *believe* H either, but it is easy to adapt his response to defend an inductivist approach to the underdetermination problem. Hence, it might be argued that if H has previously been predictively successful, and G is ad hoc in the sense of being introduced merely to accommodate the data without entailing any new predictions, then, given the past success of the overall method of believing empirically successful theories over ad hoc ones, we have inductive grounds for thinking H and not G is likely to be true (compare 2.2, pp. 49–50).

For example, suppose that H is the theory that all life on Earth has evolved by a process of random mutation and variation and natural selection in favour of variants that are fit within their niche. This hypothesis entails that human beings evolved from other species, and since the closest species to us are primates like chimpanzees, it suggests that humans and chimpanzees have some relatively recent common ancestors. So, together with background assumptions, H implies that we will find evidence of species that are intermediate in morphology between humans and primates. This is just what happened when people began to unearth the skulls and bones of so-called hominids such as *Australopithecus* and so on. Now suppose G is the hypothesis that God created the world six thousand years ago with all the supposed evidence of evolution ready made. G is certainly consistent with the evidence we have gathered, but it entails no new predictions and seems ad hoc. In practice, it seems that these considerations are sufficient to persuade most scientists to accept the evolutionary hypothesis instead of the creationist alternative.

The weak underdetermination argument is a form of the problem of induction; if H is any empirical law, such as all metals expand when heated, G states that everything observed *so far* is consistent with H but that the next observation will be different. Again, many philosophers deny the second premise. In other words, they argue that the mere fact that everything observed so far is logically consistent with metals suddenly behaving differently does not mean we have no reason to believe that metals will continue to expand tomorrow. We have already seen that the problem of induction challenges any positive account of scientific methodology and knowledge, whether or not we are talking about theories that posit unobservable entities. Let us assume that the problem of induction has a solution. There are other underdetermination arguments based on the schema above. The argument can be further strengthened by changing (2) above so that it is claimed that the alternative hypothesis is consistent, not just with all the data we have *actually* gathered, but with *all* the predictions that the other theory entails. This makes (3) more plausible, because if it is the case that all the predictions of H and G for what we can observe are the same, then it is harder to see what could justify our preference for H.

6.1.2   Strong underdetermination

The appeal to strong underdetermination is a general sceptical form of argument that is often used in philosophy. Descartes' dream argument in his first meditation is a good example (1641). He argues as follows:

(a) Sometimes dreams can be very realistic so that *during* the dream it is indistinguishable from everyday life.
(b) If you cannot distinguish being awake from dreaming then it is possible that you are dreaming now.
(c) If you cannot eliminate the possibility that you are dreaming now, then you don't know that you are not dreaming now.

Therefore you don't know you are not dreaming now.

Other examples of strong underdetermination arguments that we have already discussed include Hume's argument against a necessary connection between things we call causes and effects, and Descartes' evil demon (or the brain in a vat) argument against our knowledge of the external world. In each case, it is argued that all the evidence we could ever have is not sufficient to rule out an alternative hypothesis, and then that we cannot know the thing we take ourselves to know, if we cannot know that this alternative hypothesis is false:

(1) We think we know p (there are necessary connections/the external world I see around me exists).
(2) If we know p then we must know that q (there are no necessary connections / I am a brain in a vat) is false (because p implies that q is false).
(3) We can't know q is false.

Therefore we don't know p after all.

Hence, scepticism about some belief can be motivated by finding a rival hypothesis that predicts that everything will *seem* the same. It is easy to think of other hypotheses to fit this schema. For example, all the data we now have is consistent with the theory that the world came into existence five minutes ago, with all our memories and the physical records of the past as part of it.

Those who argue this way support (3) by saying that we cannot know that the alternative hypothesis q is false, because everything we can observe is consistent with q. The idea of indistinguishability is crucial here. Suppose someone claims the bird they have just seen is an eagle, but when asked how they know, he or she cites the evidence of talons and a curved beak. There are other birds of prey that have talons and a curved beak, but which are not eagles. If someone cannot distinguish an eagle from a buzzard, then, intuitively, he or she does not know it is an eagle they are seeing. Clearly the possibility that the bird is a buzzard is *relevant* in the circumstances. Perhaps, the reason we do not consider the hypothesis that we are all brains in vats is because it is not such a relevant alternative. Then we could argue that the second premise above is false, because it is not necessary for knowing p that we can rule out q, if q is not a relevant alternative. After all, surely I do not have to be able to rule out the possibility of random teleportation or spontaneous combustion to know that there is bread in the cupboard.

Is the possibility that you are dreaming now relevant to whether you know you are reading these words? Perhaps, but even if it is possible to have a dream that is indistinguishable from mundane waking life while it is happening, it is obvious that for most of us our waking life has a coherence that our dreams do not. Days follow one from another, and the effects of what we did yesterday are seen today. What about Descartes' evil demon, or the thought that you might be a brain in a vat being fed what you see and hear by a supercomputer? In ordinary life, such possibilities are not relevant, but when we are reflecting on what we really know the doubts they raise are harder to dispel. However, suppose that it is quite reasonable to reject the bizarre thought that the world we see around us might not exist. Hume's argument against a necessary connection between cause and effect is still interesting because it challenges a specific metaphysical doctrine about causation. Many argue that where everything we can observe supports two hypotheses equally, then we ought to adopt the metaphysically less extravagant hypothesis (see the discussion of causation in 2.1). Once we reflect on the nature of causation, Hume's alternative theory, according to which there is nothing in the world corresponding to our idea of causation other than regularities in phenomena, is surely relevant and is certainly less metaphysically extravagant, than the view that there are necessary connections of some kind. Of course, as we saw in the previous chapter, positivists seek to rule out as meaningless any hypothesis the truth or falsity of which, in principle, cannot be decided by experience. It is because they think that all the possible data underdetermine which of the following is correct, that positivists say they are merely metaphysical nonsense:

> Everything that happens is a random result of physical forces.
> Everything that happens is designed for a reason by God.
> Everything that happens is an effect of a previous cause.

If scientific realists are claiming knowledge beyond empirical facts, then, like theologians and the philosopher who believes in necessary connections, they may be vulnerable to a strong form of under-determination argument. It is usually motivated by appeal to the Duhem–Quine thesis.

6.1.3   The Duhem–Quine thesis

In 3.4 I explained the Duhem problem, which is that, because several hypotheses as well as initial and background conditions are always needed to derive any predictions, crucial experiments between rival scientific theories can never show as a matter of logic that one particular hypothesis is true or false. Hence, only theoretical systems as a whole can be tested against experiment (this is called confirmational holism). This means that when we are testing two theories, say the wave theory of light and the particle theory of light, and some phenomenon is observed that seems to falsify one of them, the falsification may be located with one of the auxiliary or background assumptions by those who wish to hang on to their theory. Then they may attempt to modify some of their background assumptions to solve the problem. (Sometimes this turns out to work, as we saw in 3.4.) The problem is to give a principled account, which can be applied without the benefit of hindsight, to tell us when it is justified to retain a theory in the face of apparent refutation, and when it is not.

Duhem argued that, in practice, scientists' 'good sense' is necessary to locate the source of falsification within the theoretical system. This does nothing more than to restate the problem because Duhem does not explain what principles good sense is based on, nor how exactly it works. This presents a difficulty for any account of scientific methodology, but it may not be so intractable. For one thing, the auxiliary assumptions corresponding to initial conditions and specific facts about the particular experimental context may be varied and the experiment repeated, and if the expected observation still fails to occur then scientists may be sure that it is not those that are to blame. This is why scientists try and repeat important experiments in different laboratories, so that, for example, effects that may be artefacts of particular equipment faults or human errors are eliminated. Furthermore, once again it may be argued that inductive support weighs against auxiliary assumptions that do not entail new predictions and that are only introduced to save a theory from refutation. In any case, this is still a weak form of underdetermination to be solved by the evolving practice of science.

As I mentioned in 3.4, Quine's holism is more radical because he argues that the Duhem problem can be extended so that mathematics and logic are included in the theoretical system, and that it may be reasonable to regard an experiment as falsifying them rather than the empirical part of the system: 'any statement can be held true come what may, if we make drastic enough adjustments elsewhere in the system' (Quine 1953: 43). Hence, he argues that the smallest thing that has empirical significance is the *whole* of science, which includes logic and mathematics as well. This suggests a very strong form of underdetermination. If Quine is right, experiment underdetermines whether we should regard a particular theory as falsified or confirmed, or instead take the experiment as showing us that arithmetic is wrong. If the same is true of logic then we are really in trouble. Consider the following.

Some theory, T, together with background assumptions, B, implies that some phenomenon, e, will be observed. (It could be a theory of particle physics, which says that electrons decay into other particles at a certain rate, with background assumptions about the likely types of particles produced, their masses and so on. In this case, e would be one of a class of characteristic traces of collisions in a particle accelerator.) Suppose e is not detected. The scientists deduce that either the theory or one of the background assumptions is false. The background theories, and the procedures for determining the initial conditions, are all well tested and used regularly in other contexts, and so using their good sense, they assume that T is false. (This is in fact the case, and most physicists think that electrons are truly elementary particles, in the sense of not being composed of further parts.) The familiar Duhem problem challenges us to explain exactly how they know what to regard as falsified. Maybe electrons do decay after all but even more rarely than T predicts, or maybe one of the background theories is false after all. However, even if that remains mysterious, at least it seems we know that one of T and B is false. The inference has this form:

$T\&B{\rightarrow}e$ (B represents $B1\&B2\&\ldots\&Bn$, where the Bs are the
              n background assumptions)

Not e

Therefore, not ($T\&B$)

Intuitively, this is a correct form of argument. If p implies q, and q is false, then p cannot be true (for if it was, q would be as well and it isn't). However, this is true if p and q are any propositions at all, so we can set p to be $T\&B$, and q to be e. It is also a law of logic that not ($T\&B$), which says that it is not the case that T and B are both true, is equivalent to 'not T or not B', which says that either the negation of T is true or the negation of B is true, therefore it seems that if e does not happen then either T or B (or both) must be false. According to

Quine's radical confirmation holism, instead of assuming that either T or B (or both) is false, we could take the falsity of e to show that one or more of the rules of logic we have used here are wrong.

Quine thinks, in practice, we rely upon pragmatic considerations to solve this extreme underdetermination problem. So, for example, he thinks that it would be so inconvenient to change the laws of logic in normal applications, that it will always be easier to change other parts of the system. Mathematics is a more difficult case. One reason why radical confirmational holism is worth discussing is that there is a historical example where the evidence that supports one physical theory has been taken as thereby falsifying a part of mathematics. Euclidean geometry, the mathematics of points, lines and angles, was once regarded as an *a priori* science of the structure of physical space. Kant thought that, since the theorems of Euclidean geometry follow deductively from axioms that seem self-evidently true (such as that between any two points there is a straight line joining them), and because theorems like 'all triangles have angles totalling to 180 degrees' seem true of real triangles, Euclidean geometry was not only *a priori*, but also synthetic (see 5.3.1); in other words, substantial knowledge of the world known by the use of reason alone. It is hard to overestimate the importance of Euclidean geometry in the history of mathematics, science and philosophy. For a long time before Kant, it was a model for the structure of scientific theories, and both Descartes and Newton, among others, tried to emulate its style.

The questioning of Euclidean geometry began in the nineteenth century when two alternative systems of geometry were shown to be consistent. Both of these systems denied the famous fifth and final axiom of Euclid's theory, namely that given any line, l, and a point, p, not lying on the line, there is one and only one line through that point that is parallel to the original line. In Riemannian geometry there may be no straight line through p that is parallel to l, and in Bolyai–Lobachevsky geometry there may be more than one. Roughly speaking, this is because both these geometries are curved, in the sense that the shortest distance between two points is a curve. This is analogous to the geometry of the surface of a sphere, where the shortest distances between points are the great circles of the sphere, and where the angles of a triangle do not add up to 180 degrees. The study of these strange geometries was of great interest to mathematicians, but the discovery's full significance only became clear when general relativity became accepted by most physicists as a replacement for the Newtonian theory of space and time and gravity.

General relativity is formulated in terms of Riemannian geometry, so that the geometry of physical space (and time) is curved rather than flat. If the most empirically adequate physics does not employ Euclidean geometry, there is no longer much reason to regard Euclidean geometry as *a priori* knowledge of space. Many argue that this episode teaches us that there is a difference between pure and applied geometry. Pure geometry is the study of different mathematical systems described by different axiom systems, some of which have many or even infinite dimensions. It may be *a priori*, but it is abstract and nothing to do with the physical world. Applied geometry is the application of mathematics to reality. How this is to be done is learned by experience, and any knowledge we might have of the geometry of physical space is therefore empirical, not *a priori*. Some realists then argue that experience has taught us that space and time really have a geometry that is non-Euclidean. According to general relativity, light travels along the shortest lines between points, but sometimes these lines are curves because the presence of matter deforms spacetime. Gravity is not a force but the name we give to the effect of the curvature of spacetime on the motion of objects.

Of course, we have yet to decide whether we ought to be realists; perhaps general relativity is just the best guide we have to what we can observe. However, either way, we now have greater resources from which to construct strongly empirically equivalent rivals to a given theory, because we can vary the mathematics that contributes to the theoretical system. A canonical form of empirical equivalence for spacetime theories that employs such a method was presented by the great mathematician and philosopher Henri Poincaré (1854–1912, see Poincaré 1905: 65–68). The idea is that we cannot decide by experiment whether or not the world is Euclidean or non-Euclidean in its geometric structure, because empirical equivalence with any non-Euclidean theory can always be achieved by maintaining Euclidean geometry, but adding forces acting on all bodies in such a way as to mimic the effect of the non-Euclidean structure. The forces affect our measuring rods and clocks so as to create the phenomena consistent with a curved spacetime. This can be done the other way around; hence, there is a theory that is empirically equivalent to Newtonian mechanics, but which employs a curved spacetime to mimic the effects of Newton's gravitational force. However, the theories we get using this method may be completely ad hoc and hideously complicated. Is that sufficient reason for us to reject them? We shall return to this issue below.

So, to generate a strong underdetermination problem for scientific theories, we start with a theory H, and generate another theory G, such that H and G have the same empirical consequences, not just for what we have observed so far, but also for any possible observations we could make. If there are always such *strongly empirically equivalent* alternatives to any given theory, then this might be a serious problem for scientific realism. The relative credibility of two such theories cannot be decided by any observations even in the future and therefore, it is argued, theory choice would be underdetermined by all possible evidence. If all the evidence we could possibly gather would not be enough to discriminate between a multiplicity of different theories, then we could not have any rational grounds for believing in the theoretical entities and approximate truth of any particular theory. Hence, scientific realism would be undermined.

The *strong form of the underdetermination argument* for scientific theories is as follows:

(i)   For every theory there exist an infinite number of strongly empirically equivalent but incompatible rival theories.

(ii) If two theories are strongly empirically equivalent then they are *evidentially equivalent*.

(iii) No evidence can ever support a unique theory more than its strongly empirically equivalent rivals, and theory-choice is therefore radically underdetermined.

This argument is clearly valid if we assume that theory choice must be based on evidence. In what follows various strategies for avoiding the sceptical conclusion of this argument will be assessed.

6.1.4 Responses to the strong underdetermination argument

1. The strong empirical equivalence thesis ((i) above) is incoherent.
2. The strong empirical equivalence thesis is false.
3. Empirical equivalence does not imply evidential equivalence ((ii) is false).
4. Theory choice is underdetermined ((iii) is true).
   4.1 Reductionism
   4.2 Conventionalism
   4.3 Antirealism

(1) The alleged incoherence of the strong empirical equivalence thesis

There are various ways of arguing that strong empirical equivalence is incoherent, or at least ill-defined:

(a) The idea of empirical equivalence requires it to be possible to circumscribe clearly the observable consequences of a theory. However, there is no non-arbitrary distinction between the observable and unobservable.

The most obvious way to undermine (i) above is first to point out that even to articulate the notion of the empirical consequences of a theory is to make principled use of the distinction between the observable and the unobservable. It may be then argued that there is no sharp way to draw such a distinction, and that it is therefore arbitrary and cannot bear the epistemological weight put upon it. The point is that there is a grey area between what is clearly observable, like a tree, and what is clearly unobservable, like an electron. So, for example, whether things we can only see with a magnifying glass, like amoeba and tiny insects, are to be counted as observable or not seems not to be well-defined. (We shall return to this issue in section 6.2.1 below.)

This objection is inconclusive for two reasons. First, the distinction need not be sharp to be non-arbitrary. There may be cases where it is not clear whether we have observed something or not, or whether some of the consequences of a theory are observable or not.

175

However, as long as there are still plenty of other definite cases, then the empirical consequences of a theory will be clear enough for the purposes of the argument. It is obvious that electrons and quarks (which are supposed to compose protons and neutrons and hence all atomic nuclei) are not observable, and that rivers, people and everyday objects are observable. Secondly, the observable/unobservable distinction may be made sharp in an epistemologically well-motivated way for the purposes of identifying observational consequences of specific theories. One way of doing this is to identify the observable terms of a theory with those which are understood independently of that particular theory. Hence, terms such as velocity, momentum, temperature and position are observable terms with respect to electromagnetic theory, whereas terms such as electromagnetic potential, and charge density are not.

(b) The observable/unobservable distinction changes over time, and so what the empirical consequences of a theory are is relative to a particular point in time.

The second response to (a) implicitly concedes that the observable/unobservable distinction will change over time, because our theories change over time. Recently, Larry Laudan and Jarrett Leplin have argued that distinguishing the empirical consequences of a theory requires identifying 'those of its logical consequences formulable in an observation language' (Laudan and Leplin 1991: 451), and that the observation language is not fixed: 'Any circumspection of the range of observable phenomena is relative to the state of scientific knowledge and the technological resources available for observation and detection' (Laudan and Leplin 1991: 451). So whether or not two theories are empirically equivalent will change with time. Therefore, they argue, although theories may be empirically equivalent momentarily they are not eternally so, and hence we have the chance to distinguish them in the future. They think this defeats even the strong underdetermination argument.

However, there is an obvious rejoinder, namely, that we now have a relativised notion of empirical equivalence. Theories can still be empirically equivalent at a time, so we have *synchronic* but not *diachronic* empirical equivalence. We can then simply relativise the underdetermination argument so that it applies to theories at particular times. This still means that, at any given time, given any scientific theory, there will be a theory that is empirically equivalent to it at that time, and so theory choice will be underdetermined at that time. Another response to (b) is that of Hoefer and Rosenberg (1994). They argue that the scope of (i) should be restricted to 'total' (as opposed to 'partial') theories; that is, ones that are globally empirically adequate in the sense that they predict *all* the phenomena, not just those in one area of science. If we had such a theory then we would have a privileged account of what is observable. It is surely a problem for the scientific realist, if even a total theory would have empirically equivalent but incompatible rivals.

(c) Theories only have empirical consequences relative to auxiliary assumptions and background conditions. So the idea of the empirical consequences of the theory itself is incoherent.

The Duhem–Quine problem arises because drawing specific empirical consequences from some theory always requires the addition of initial values of certain quantities, values for physical constants (such as the gravitational constant or the mass of the electron) and so on. Usually, a theory must be combined with background theories from other parts of science to determine most of its empirical content. Hence, it is claimed that whether or not two theories are empirically equivalent depends on what auxiliary hypotheses are brought into play. So, in the case of alleged empirical equivalence it may be possible to distinguish between the theories if we can identify suitable auxiliary or background assumptions. Furthermore, there is also a diachronic version of this argument which says that empirical equivalence holds not between theories *per se* but only between them when conjoined with certain other hypotheses at a particular time (cf. Laudan and Leplin 1991). Once more there is the possibility, or even the probability, that in the future when new auxiliaries are chosen the theories will not after all be empirically equivalent.

This is an important objection to the idea of empirical equivalence as it is deployed in the underdetermination argument. However, (i) can be reformulated so that it refers to empirically equivalent total theories, where the latter are self-contained with respect to auxiliary hypotheses. Corresponding to their distinction between partial and total theories, Hoefer and Rosenberg make a distinction between 'local' and 'global' underdetermination. Obviously, if a theory is only partial, and so does not account for some range of phenomena, then it is possible, indeed probable, that its empirically equivalent rivals will not be empirically equivalent to it when it has been extended to deal with other phenomena. So partial theories will only be weakly underdetermined. We saw above that this does not pose a special problem for scientific realism. The underdetermination problem is worse for realists only if we are talking about the strong empirical equivalence of total sciences (or 'systems of the world' as Hoefer and Rosenberg call them). So the question is whether there are any examples of total theories that have strong empirical equivalents.

(2) The alleged falsity of the strong empirical equivalence thesis

It may be argued that there is no reason to believe that there will always exist strongly empirically equivalent rivals to any given theory, either because cases of strong empirical equivalence are too rare, or because the only strongly empirically equivalent rivals available are not genuine theories.

The scientific realist is not committed to the view that all the best current scientific theories ought to be believed but, rather, that there are at least some circumstances in which we ought to believe in the truth and successful reference of theories about the unobservable. So to establish (i) it must be shown that empirically equivalent rival theories always exist, not just that they do in some cases. Of course, we can always generate a theory that is empirically equivalent to some theory, T, by simply conjoining propositions to T that entail nothing else of an empirical nature. For example, the mini-theory 'electrons are charged and God exists' implies everything that the theory 'electrons are charged' does (for example, that electrons will be deflected in some types of magnetic field), assuming we combine them with the same background theories. In this way we do not generate a theory that is incompatible with T, but there are other algorithms for generating empirically equivalent rivals. A trivial one is as follows: given any theory T, let T' be the assertion the empirical predictions of T are all true, but the theoretical entities posited by T

do not exist. Here, T and T' are incompatible but have the same empirical consequences by definition. Andre Kukla gives another example of an algorithm that produces a rival to any theory T:

> For any theory T, construct the theory $T!$ which asserts that T holds whenever somebody is observing something, but that when there is no observation going on, the universe behaves in accordance with some other theory T_2 which is incompatible with T.
>
> (Kukla 1998: 70)

(The reference to T in $T!$ is eliminable, because we can replace the name 'T' with the specific laws it asserts.)

The critic of the underdetermination argument may argue that such algorithms fail because they produce pseudo-theories. Hence, Laudan and Leplin call such algorithms 'logico-semantic trickery' (Laudan and Leplin 1991: 463), and Hoefer and Rosenberg call them 'cheap tricks' (Hoefer and Rosenberg 1994: 604). They all argue that such empirically equivalent theories are not to be counted among the rivals of a theory for the purposes of deciding whether (i) is true or not. However, prima facie, these theories ought to be included. After all, they are assertible and have a truth value determined by truth conditions like those for any other putative truth about the world, and they have empirical import making them as amenable to testing as any other theory. The algorithms above may produce theories that are artificial, but we need to know what the criteria are for being a 'proper' theory that will distinguish the products of such algorithms from the original theory upon which they act, in such a way that we can see why the evidence supports proper theories more. Hoefer and Rosenberg concede that if the world is sufficiently complex, it is possible that the globally empirically adequate theory would have to be a list of all the actual observable phenomena that occur, because there just may not be any more concise theory that saves all the phenomena (Hoefer and Rosenberg 1994: 605), so we cannot reject theories just because they are clumsy.

If certain theories are ruled out for the purposes of deciding whether (i) is true or not, because they are artificial and complex, this amounts to the claim that there are non-empirical, but nonetheless rational, grounds for preferring one theory to another, whether because of simplicity, 'intrinsic plausibility' (Horwich 1991), or whatever. (As Kukla puts it, 'theoreticity' – being a proper theory – is being adopted as a superempirical virtue of theories, see Kukla 1996: 147). Hence, it seems this way of disputing the truth of premise (i) is really a way of disputing premise (ii), because if there are non-empirical grounds for believing theories to be true or more likely to be true than their rivals, then the empirical equivalence of theories will not imply their evidential equivalence. We shall return to this issue in the next section.

Even if T' and $T!$ are genuine alternatives to T, are they relevant alternatives? It certainly seems that scientists do not regard them as relevant. What we really need is a theory that seems to be a proper scientific alternative. Instead of the philosophers' tricks used above there are more interesting ways of constructing empirically equivalent theories, for instance, by replacing the unobservable structures of a theory with extra 'self-correcting' structures which yield exactly the same observational predictions. As we saw above, this is the method used by Poincaré in his spacetime example. Similar cases arise in contemporary physics where there are several competing versions of quantum theory. However, in many of these cases the theories produced are as artificial and ad hoc as those produced by philosophers' tricks, so they too may not count as genuine and relevant empirical equivalents.

However, there are cases where we do not need to construct a new theory to generate empirical equivalents, because if a theory's predictions are independent of the value of some parameter, then we can generate empirical equivalents by varying that parameter. A famous example of this kind of empirical equivalence is afforded by Newtonian mechanics. Let Newton's laws of motion plus the law of gravitation be TN, then call this theory plus the hypothesis that the centre of mass of the solar system has constant absolute velocity v the theory $TN(v)$. Since absolute motions are undetectable according to TN (the theory is said to be Galilean invariant), then $TN(0)$ has all the same empirical consequences as $TN(v)$, for all v. However, these are clearly incompatible theories and we therefore have an example of strong empirical equivalence. Hoefer and Rosenberg dismiss this example because, since TN cannot account for the phenomena predicted by general relativity, it is not empirically adequate. They argue that we can conclude nothing about whether or not a globally empirically adequate theory would have empirically equivalent rivals, from examples of theories that are known to be false and empirically inadequate. The same goes for general relativity, because it cannot account for forces other than gravity, and for quantum theory, because it cannot account for gravity.

In conclusion, it seems that interesting examples of strong empirical equivalence have been found for putative global theories that are, in fact, empirically false. However, we have no grounds for thinking that a global theory that *is* empirically adequate will have strong empirically equivalent rivals, other than those produced by logical, semantic, or mathematical tricks. Hence, although (i) has not been shown to be false for global theories, neither has it been shown to be true. The plausibility of the underdetermination argument depends on whether we can rule out the artificial theories of philosophers, and more generally on the fate of premise (ii).

(3) Empirical equivalence does not imply evidential equivalence ((ii) is false)

Whether or not any of the objections to (i) we have considered so far work, many realists argue that (ii) is false. They argue that two theories may predict all the same phenomena, but have different degrees of evidential support. In other words, they think that there are non-empirical features (*superempirical* virtues) of theories such as simplicity, non-ad hocness, novel predictive power, elegance, and explanatory power, that give us a reason to chose one among the empirically equivalent rivals.

Scientific theories can be said to have 'empirical virtues' such as empirical adequacy and empirical strength. A theory is empirically adequate if what it says about the phenomena or the observable facts is true. So, for example, the theory that all copper expands when heated is empirically adequate. Theories may be more or less empirically adequate. Of course, theories that make quantitative predictions can be approximately empirically adequate. For example, general relativity is more empirically adequate than Newtonian theory, because it gets the phenomena, such as the orbit of Mercury, right to a greater degree of accuracy. However, two theories may have differing degrees of empirical strength even though the weaker one is empirically adequate as far as it goes. For example, the theory that all metals expand when heated is empirically stronger than the theory that all copper expands when heated, because it entails more predictions, or has more scope. It is obvious that part of the aim of science is to produce empirically adequate and strong theories; if we want to know about the world then scientific theories need to be consistent with the facts, and the more a particular theory can tell us the better. However, are there any other virtues that theories have which are also relevant to theory choice?

It certainly seems to be the case that, in practice, scientists do not choose among theories on purely empirical grounds. There are many cases in the history of science where eminent scientists have justified their preference for a particular theory by citing its simplicity, explanatory power, coherence with other parts of science, or coherence with background metaphysical views such as materialism. It is also common for scientists to talk about theories being beautiful, or elegant, and to suggest that the possession of such features influences theory-choice. They often prefer a theory $T1$ to its locally empirically equivalent counterpart $T2$, because $T1$, but not $T2$, is embedded in a more comprehensive theory T^* that has independent support. For example, the kinetic theory of gases, which is the theory according to which heat and other thermodynamic properties of gases are caused by the motions of the molecules that make them up, was preferred to its rivals because it could be embedded in atomic theory, which is also supported by evidence derived from chemistry and other parts of physics. In general, it seems that superempirical virtues are what scientists use to solve local underdetermination problems.

However, it is a problem for the realist that there is no agreed way to rank these virtues, nor agreement about how to proceed when they pull in different directions. For example, the special theory of relativity offered a simple and unifying account of the motion of charged particles in electromagnetic fields, but it entailed a radical revision of the Newtonian ideas of mass, space and time. So it seems to have superempirical virtues and vices. No scientific theory has ever been unified with every other theory, or compatible with all background metaphysical views. It is also difficult to explain why a simple or elegant theory is more likely to be true than a complicated and ugly one. Of course, we still have good reason to try and get simple theories, because there are practical considerations about the time it takes to do calculations and remember lots of very complicated equations. It is easier to have an equation that gives the positions of the planets than it is to have an enormous list of all the positions of all the planets at every second. Similarly, the antirealist can explain the fact that most of the most successful scientific theories we have (such as Newton's mechanics, Maxwell's electrodynamics, and the theory of chemical bonding) have a few basic equations and principles at their core, by saying that this is a practical necessity, because the difficulty in applying such laws to messy real-world situations would be insurmountable if the fundamental theory was itself vast and complex.

Hence, van Fraassen argues that the superempirical virtues do not give us reason for belief (they are not epistemic), but merely reason to adopt a theory for practical purposes (they are pragmatic). He says 'pragmatic virtues do not give us any reason over and above the evidence of empirical data, for thinking that a theory is true' (van Fraassen 1980: 4). Note that this is why it is significant that pragmatic theories of truth were ruled out of the definition of scientific realism at the end of the previous chapter. To put it crudely, for someone who holds a pragmatic theory of truth there is no difference between pragmatic and epistemic virtues because there is no difference between what it is most useful to believe and what is true.

So, to defend realism, it is insufficient to appeal to the use of superempirical virtues to solve underdetermination in the practice of science without explaining why we cannot treat them as pragmatic rather than epistemic. As with so many of these issues, a full examination of them would require us to look in detail at real cases of theory change where superempirical virtues seem to be important. In any case, the most plausible defence of the epistemic status of superempirical virtues is that which subsumes them within the overall virtue of explanatory power. On this account, simplicity, unifying power, elegance and so on are reasons for belief only in so far as they contribute to a theory being explanatory of a wide range of phenomena. This is the subject of the next chapter.

(4) Theory choice is underdetermined ((iii) is true)

There are various options if we accept the conclusion of the under-determination argument.

(4.1) Reductionism

Should a theory written in French and a theory written in English be regarded as different theories? How are we to draw the line between genuine difference in content and mere notational difference? Many empiricists have argued that where two theories are observationally equivalent, they should be regarded as different formulations of the same theory. Hence, as we saw in 5.3.2, the positivists sought to reduce theories to their observational basis.

(4.2) Conventionalism

Conventionalism is the idea that the choice between observationally equivalent theories is a convention. Just as it doesn't matter whether we drive on the left or the right, as long as everyone follows the same convention, so the conventionalist argues that the choice between empirically equivalent theories is to be made on the basis of convenience, or perhaps as a result of features of our cognition, such as the way visual perception or language work. This option is clearly not available to the realist.

(4.3) Antirealism

Another response to underdetermination is to adopt some form of social constructivism about scientific knowledge. Hence, it is often argued by those influenced by Kuhn and the sociology of science (see (4.7)) that the underdetermination of theories is broken not by superempirical virtues, but by social, psychological and ideological factors. Some scientists believe in an order to nature created by God, some want to defend materialism, some just want a theory that works because it is needed for technology that is profitable or socially important, some want to understand the inner workings of nature, and so on. Most realists do not deny that such factors have an effect on the practice of science, but argue that it is swamped by the rigours of repeated experimental testing, peer review, and scientists' commitment to be open minded and sceptical about all theories.

Social constructivists deny the metaphysical component of scientific realism. Another response to the underdetermination argument is to deny the epistemic component. This may take the form of arguing that our best theories are most likely false (atheism), or simply suspending judgement about the truth of scientific theories and the nature of the unobservable world (agnosticism). The latter type of antirealism is the subject of the rest of this chapter.

6.2 Constructive empiricism

The constructive empiricism of van Fraassen has provoked renewed debate about scientific realism. Van Fraassen is happy to accept the semantic and metaphysical components of scientific realism that I explained in Chapter 5, but he denies the epistemic component. He thinks that scientific theories about unobservables should be taken literally, and are true or false in the correspondence sense, depending on whether the entities they describe are part of the mind-independent world. However, he argues that acceptance of the best theories in modern science does not require belief in the entities postulated by them, and that the nature and success of modern science relative to its aims can be understood without invoking the existence of such entities.

Van Fraassen defines scientific realism as follows: 'Science aims to give us, in its theories, a literally true story of what the world is like; and acceptance of a scientific theory involves the belief that it is true' (van Fraassen 1980: 8). On the other hand, constructive empiricism is the view that: 'Science aims to give us theories which are empirically adequate; and acceptance of a theory involves as belief only that it is empirically adequate' (van Fraassen 1980: 12). To say that a theory is empirically adequate is to say: 'What it says about the observable things and events in this world, is true' (van Fraassen 1980: 12). In other words: 'the belief involved in accepting a scientific theory is only that it "saves the phenomena", that is that it correctly describes what is observable' (van Fraassen 1980: 4). Note that this means that it saves *all* the *actual* phenomena, past present and future, not just those that have been observed so far, so even to accept a theory as empirically adequate is to believe something more than is logically implied by the data (van Fraassen 1980: 12, 72). Moreover, for van Fraassen, a phenomenon is simply an *observable* event and not necessarily an observed one. So a tree falling over in a forest is a phenomenon whether or not someone actually witnesses it.

The scientific realist and the constructive empiricist disagree about the purpose of the scientific enterprise: the former thinks that it aims at truth with respect to the unobservable processes and entities that *explain* the observable phenomena; the latter thinks that the aim is merely to tell the truth about what is observable, and rejects the demand for explanation of all regularities in what we observe. Van Fraassen says that explanatory power is not a 'rock bottom virtue' of scientific theories whereas consistency with the phenomena is (van Fraassen 1980: 94). Hence, for the constructive empiricist, empirical adequacy is the internal criterion of success for scientific activity.

6.2.1 Objections to constructive empiricism

The most common criticisms of constructive empiricism are the following:

(i) The line between the observable and the unobservable is vague and the two domains are continuous with one another; moreover, the line between the observable and the unobservable changes with time and is an artefact of accidents of human physiology and technology. Hence, it is argued that constructive empiricism grants ontological significance to an arbitrary distinction.

(ii) Van Fraassen rejects the positivist project that attempted to give an *a priori* demarcation of terms that refer to observables from those that refer to unobservables, and accepts instead that: (a) all language is theory-laden to some extent; (b) even the observable world is described using terms that putatively refer to unobservables; and (c) acceptance of a theory involves a commitment to interpret and talk about the world in its terms. Critics argue that this makes van Fraassen's position incoherent.

(iii) The underdetermination of theory by evidence is the only positive argument for adopting constructive empiricism instead of scientific realism; but all the data we presently have underdetermine which theory is empirically adequate (the problem of induction), just as they underdetermine which theory is true, and so constructive empiricism is just as vulnerable to scepticism as scientific realism. This is taken to imply that van Fraassen's advocacy of constructive empiricism is the expression of an arbitrarily selective scepticism.

(1) The observable and unobservable

The first and most fundamental realist objection to constructive empiricism is that no meaningful line can be drawn between the observable and the unobservable, and the second is that even if such a demarcation is possible, there are no grounds for thinking that it has any ontological or epistemological significance. This realist intuition that there is nothing special about unobservable entities that prevents them existing (or us knowing about them existing) is well put by Grover Maxwell, who argued that there is a continuum between seeing normally, seeing through a window, seeing with binoculars, seeing with a microscope, and so on, such that we cannot even draw the line between the observable and the unobservable apart from in an arbitrary way. Hence, he said:

> [The] drawing of the observational theoretical line at any given point is an accident and a function of our physiological make-up, our current state of knowledge, and the instruments we happen to have available and, therefore, it has no ontological status whatsoever.
>
> (Maxwell 1962: 14–15)

In other words, why should the boundary between what exists and what does not fall where we happen to place the boundary between what we are able to observe and what we cannot?

Van Fraassen agrees that this need not be the case, and so concedes that unobservable entities *may* exist; however, he does think the boundary between what we can and cannot *know* exists coincides with the boundary between the observable and unobservable: 'even if observability has nothing to do with existence (is, indeed, too anthropocentric for that), it may still have much to do with the proper epistemic attitude to science' (van Fraassen 1980: 19). Hence, van Fraassen's antirealism is epistemological, not metaphysical. In response to the claim that the continuum between aided and unaided acts of perception prevents any distinction being drawn between the observable and unobservable, van Fraassen points out that almost all predicates, like 'is red', 'is a mountain' and so on, are vague at the boundaries, and says that, as long as there are clear extreme cases, this does not prevent us using them.

Let us suppose that there do exist clear extreme cases that can be unambiguously classified as observations or not, it is still not clear how the possibility of drawing such a distinction can support scepticism about the so-called unobservable. Any act of perception may be an observation or not, but this does not amount to showing that the *objects* of perception can be classified as *observable* or not. Indeed, Maxwell argues that in fact *nothing* is 'unobservable in principle', on the grounds that this could only mean that the relevant scientific theory entailed that the entities could not be observed in any circumstances; this is never the case since the different circumstances could involve our having different senses. Suppose, for example, we were like aliens with electron microscopes for eyes; atoms *would* be observable (see Churchland 1985).

However, for van Fraassen, 'observable' is to be understood as 'observable-to-us': 'X is observable if there are circumstances which are such that, if X is present to us under those circumstances, then we observe it' (van Fraassen 1980: 16). We may ask who the relevant 'us' is, does it include those of us who are partially sighted or blind? If so then how can things be unambiguously observable or not? Van Fraassen's response is that 'us' refers to our epistemic community, and that this includes the partially sighted, and the eagle sighted who can spot dim stars that most of us cannot. What we can and cannot observe is a consequence of the fact that:

The human organism is, from the point of view of physics, a certain kind of measuring apparatus. As such it has certain inherent limitations – which will be described in detail in the final physics and biology. It is these limitations to which the 'able' in 'observable' refers – our limitations, qua human beings.

(van Fraassen 1980: 17)

If this community was to change in some way or other – perhaps we do meet aliens who claim that their eyes work just like our electron microscopes and we become integrated into each other's scientific communities – then the demarcation between observable and unobservable will change appropriately. In the absence of these aliens, what is unobservable and unknowable is determined by the best of our senses.

So we know that, for example, the moons of Jupiter are observable because our current best theories say that, were astronauts to get close enough, then they *would* observe them. On the other hand, the best theories of particle physics certainly do not tell us that we are directly observing the particles in a cloud chamber. Analogous with the latter case is the observation of the vapour trail of a jet in the sky, which does not count as observing the jet itself, but rather as detecting it. Now if subatomic particles exist, as our theories say that they do, then we detect them by means of observing their tracks in cloud chambers for example, but, since we can never experience them directly (as we can with jets), there is always the possibility of an empirically equivalent but incompatible rival theory which denies that such particles exist. (Note that this means van Fraassen adopts a direct realism about perception for macroscopic objects: 'we can and do see the truth about many things: ourselves, others, trees and animals, clouds and rivers – in the immediacy of experience' (van Fraassen 1989: 178).)

The question then posed by critics is why are we allowed to imagine changing our spatio-temporal location when determining what is observable, but not our size or the configuration of our sensory apparatus? As van Fraassen says, this argument of Churchland and others has the following form:

We could be, or could become, X. If we were X, we would observe Y. In fact, we are, under certain realizable conditions, like X in all relevant respects. But what we could under realizable conditions observe is observable. Therefore, Y is observable.

(van Fraassen 1985: 257)

The third premise above need not be believed unless one already accepts that the relevant theory is true. This enables us to distinguish between the two cases above: in the case of the moons of Jupiter, we must believe that we are, in all relevant respects, like beings whose only difference is that they are closer to Jupiter; but we need not believe that we are, in all relevant respects like beings that can see electrons if we do not already know that the latter exist. The realist may sense some circularity here – do we not already have to believe that the moons of Jupiter exist in order to know that if we were closer we would see them? Not exactly, for in this case, belief that our theory of the moons of Jupiter is empirically adequate entails that if they were to be present to us we would see them; this is a disanalogy with the case of electrons – that the theory is empirically adequate says nothing about what would happen were we to be differently constituted as observers.

Since we can hold our epistemic community fixed while imagining a different spatio-temporal location for it, dinosaurs and the moons of Jupiter can be said to be observable. However, we cannot say the same of atoms, since, according to science, in order to observe them then we would have to have a different physical constitution, but we need not believe that if we had such a different physical constitution we *would* observe them, unless we already believe that they exist. Of course, the realist has contrary intuitions, and realists do not see why our physical constitution, as a contingent feature of our evolution, has any philosophical significance whatsoever. One response to this is simply to restate the opposite intuition: what else but *our* (biologically determined) observational capacities would one consider relevant to *our* epistemology?

(2) Acceptance and belief

It is important that van Fraassen agrees with realists that no distinction can be drawn between observable/unobservable terms in scientific language. Indeed, he thinks that the distinction between what is observable and what is not is to be drawn by a consideration of what our best scientific theories tell us about the entities they describe, and about our constitution as observers. It is crucial to van Fraassen's account of science that acceptance of a theory involves not just a belief in its empirical adequacy, but also: 'a commitment to the further confrontation of the new phenomena within the framework of that theory, a commitment to a research programme, and a wager that all relevant phenomena can be accounted for without giving up the theory' (van Fraassen 1980: 88).

Van Fraassen is happy to concede to the realist that we often have to use the language of science to describe the world and that this language is thoroughly theory-laden; for example, we cannot do without talking about microwave ovens and VHF receivers (van Fraassen 1980: 81). Furthermore, he admits that for many ordinary as well as scientific purposes it is necessary to immerse oneself in the world described by theories. Hence, van Fraassen accounts for these aspects of the practice of science. However, he maintains that this only ever provides pragmatic support for the theoretical commitments of the theory, and that using theoretical language and descriptions is not inconsistent with withholding belief in the truth of the theory.

One possible line of argument against constructive empiricism is to dispute that there is anything more to realism than accepting immersion in the world-picture of science. Realists have argued that constructive empiricism depends upon a substantive distinction between acceptance and belief that is simply not available. Paul Horwich (1991) has argued that the distinction between belief in the truth of a scientific theory and a supposedly weaker epistemic attitude, as recommended by van Fraassen or by instrumentalists, is incoherent. According to Horwich: 'Believing a theory is nothing over and above the mental state responsible for using it' (Horwich 1991: 2). He argues:

> If we tried to formulate a psychological theory of the nature of belief, it would be plausible to treat beliefs as states with a particular kind of causal role. This would consist in such features as generating certain predictions, prompting certain utterances,

being caused by certain observations, entering in characteristic ways into inferential relations, playing a certain part in deliberation, and so on. But that is to define belief in exactly the same way instrumentalists characterize acceptance.

(Horwich 1991: 3)

An obvious objection to this argument is that belief has the extra causal role of disposing someone to assert 'I believe theory T to be true' whereas acceptance will not dispose someone to make such assertions. However, Horwich argues that such differences in behaviour are the result not of a difference between belief and acceptance, but rather are the product of 'philosophical double-talk', so confusing people that 'they are mistaken about the right way to describe their psychological state' (Horwich 1991: 4). Hence, on his view, those who say 'I accept T but do not believe it' are wrong about what they believe, and we must explain their propensity to make such assertions in terms of factors such as their confused beliefs about belief and acceptance, and the proper attitude to science.

However, this argument seems to imply that to use Newtonian theory is to believe it, and yet this cannot be correct because many scientists use Newtonian theory every day without believing it to be true (and while believing quantum mechanics and relativity to be our most approximately true theories). Horwich also considers this objection but argues that where scientists accept a theory but do not believe it, this is always acceptance of the theory in some more or less precisely circumscribed domain (Horwich 1991: 4). Hence, he argues that we cannot make van Fraassen's style of acceptance intelligible by analogy with the common attitude of scientists who use but don't believe, for example, Newtonian theory because the former is an 'unqualified general acceptance' not a 'qualified local' acceptance (Horwich 1991: 5).

However, this is no response since we have only ever had to deal with partially empirically adequate theories anyway, and we are unlikely to be faced with a completely empirically adequate theory in the near future. Constructive empiricism is an idealisation, but to idealise here seems legitimate since the realist has just as many if not more problems with partial belief or belief in partial or approximate truth, as van Fraassen does with partial empirical adequacy. Since scientists may accept a theory in some domain without believing it, prima facie it seems that the attitude of scientists towards theories that they use but believe to be false does illustrate what is involved in acceptance that stops short of belief.

(3)   Selective scepticism?

Constructive empiricism is not the same thing as brute scepticism. Van Fraassen owes us an account of how we can have any inductive knowledge at all in the face of underdetermination. After all, why should we believe that some theory T is empirically adequate, rather than that it is merely empirically adequate until next week, or when we are looking but not otherwise? The constructive empiricist faces the underdetermination problem too, but cannot invoke explanatory power to solve it. Whether or not a consideration of explanation and inference to the best explanation is sufficient to refute constructive empiricism and establish scientific realism is the subject of the next chapter.

———o☺o———

Alice: Maybe lots of theories can all make the same predictions but that doesn't make them equally good. For one thing, some theories explain lots of different things in terms of a few basic principles. When that happens we have extra reason to think those theories are true.

Thomas: But why? Obviously it is better to have simple and unified theories but that doesn't mean they are true, just more useful.

Alice: You seem prepared to think that every success of science is just luck, but don't you see how implausible that is? If a theory manages to explain lots of different things at once then that's a good reason for thinking that it's correctly describing the world.

Thomas: I am not sure what an explanation is anyway. You know how it is with kids, every time you explain something to them they just say 'why?'. Isn't it like that with science? Nothing is ever really explained, just put in the context of more and more further facts. After all, can your book tell you why the big bang was possible?

Alice: Explanation works when people properly identify the causes of things or the laws that govern them, and the idea that science could be so successful if it wasn't getting the laws and causes of what we see around us right is crazy.

Thomas: I disagree. Science is successful because thousands of people dedicate their lives to it and lots of what they do doesn't work. It's just that we only remember the good bits.

———o☉o———

Further reading

Underdetermination

Duhem, P. (1906, tr. 1962) *The Aim and Structure of Physical Theory*, New York: Athenum.

Harding, S. (ed.) (1976) *Can Theories be Refuted? Essays on the Duhem–Quine Thesis*, Dordrecht, The Netherlands: D. Reidel.

Hoefer, C. and Rosenberg, A. (1994) 'Empirical equivalence, underdetermination, and systems of the world', *Philosophy of Science*, **61**, pp. 592–607.

Kukla, A. (1993) 'Laudan, Leplin, empirical equivalence, and underdetermination', *Analysis*, **53**, pp. 1–7.

Kukla, A. (1998) *Studies in Scientific Realism*, Oxford: Oxford University Press.

Laudan, L. and Leplin, J. 'Empirical equivalence and underdetermination', *Journal of Philosophy*, **88**, pp. 269–285.

Laudan, L. and Leplin, J. (1993) 'Determination underdeterred', *Analysis*, **53**, pp. 8–15.

Quine, W.v.O. (1953) 'Two dogmas of empiricism', in *From a Logical Point of View*, Cambridge, MA: Harvard University Press.

Sklar, L. (1974) *Space, Time and Spacetime*, Berkeley: University of California Press.

Van Fraassen, B.C. (1980) *The Scientific Image*, Chapter 2, Oxford: Oxford University Press.

Bayesian and other theories of confirmation have to offer a solution to this problem so the following all have sections on it

Glymour, C. (1980) *Theory and Evidence*, Princeton, NJ: Princeton University Press.

Horwich, P. (1982) *Probability and Evidence*, Princeton, NJ: Princeton University Press.

Howson, C. and Urbach, P. (1993) *Scientific Reasoning: The Bayesian Approach*, La Salle, IL: Open Court.

Constructive empiricism

Churchland, P. and Hooker, C. (eds) (1985) *Images of Science*, Chicago: University of Chicago Press.

Ladyman, J. (2000) 'What's really wrong with constructive empiricism?: van Fraassen and the metaphysics of modality', *British Journal for the Philosophy of Science*, **51**, pp. 837–856.

Van Fraassen, B.C. (1980) *The Scientific Image*, Oxford: Oxford University Press.

Van Fraassen, B.C. (1989) *Laws and Symmetry*, Oxford: Oxford University Press.

7

Explanation and inference

The realist places great emphasis on the power of scientific theories to explain the phenomena that they describe. Indeed, for many, explanation is a primary goal of the scientific enterprise. The basic principles of chemistry, which describe how atoms of different elements in certain fixed ratios combine to form molecules of different compounds which we can identify in the natural world (such as carbon dioxide, water and salt), are now applied in every part of science, from astrophysics to cellular biology. So most of the explanations that we have for natural phenomena involve reference to theoretical and unobservable entities. Realists argue that the truth or approximate truth of a hypothesis is a necessary condition for it to be part of a genuine scientific explanation; in other words, explanation requires more than mere empirical adequacy. Realists therefore claim that explanation in science is unintelligible from the perspective of constructive empiricism.

The superempirical virtues of simplicity, elegance, coherence with other theories or metaphysical beliefs, and so on, all seem to be desiderata for good explanations. If two theories can differ with respect to their explanatory power, even though they both predict all the same phenomena, and if explanatory power is evidence of the truth of a theory, then the empirical equivalence of theories will not imply their evidential equivalence. Much of the literature about scientific realism concerns the status of *inference to the best explanation* (hereafter IBE). This is the principle that, where we have a body of evidence and are considering several hypotheses, all of which save the phenomena, we should infer the one that is the best explanation of the evidence (providing it is at least minimally adequate according to other criteria). Realists have argued that the rule of IBE is part of the canons of rational inference and necessary for *any* substantive knowledge of the world. They go on to argue that acceptance of its reliability is all that is required to justify adoption of scientific realism, because IBE can break the underdetermination of theories.

Many defenders of IBE also argue that it is the basis of all inductive reasoning (see 2.2(6)), and hence that the constructive empiricist who rejects IBE has no basis on which to make any non-deductive inferences, even where they concern the observable, which suggests that constructive empiricism is the expression of an arbitrary selective scepticism. Hence, van Fraassen's defence of constructive empiricism must involve a critique of the status given to IBE by realists, and indeed he has marshalled a range of arguments to the effect that IBE cannot be epistemically compelling. According to him, not only can false theories provide good explanations (for example, Newtonian mechanics is false but nonetheless gives us a good explanation of the tides), but furthermore, explanatory power is a pragmatic relation between a theory, a fact and a context, where the latter is determined by the background beliefs and interests of the inquirer. Although he admits that two empirically equivalent theories may differ with respect to their explanatory power, so that explanatory considerations may break underdetermination, he argues that, since the context determines which among the scientifically relevant factors are explanatory, and since the context is relative to our interests and goals, there can be no extra epistemic support for the more explanatory theory. The search for explanatory theories is necessarily the search for empirically adequate and strong theories (because a theory that does not correctly describe what is observable cannot possibly be used to explain what we observe), and explanatory power is a purely pragmatic virtue of theories.

In the first part of this chapter we will consider the nature of scientific explanation, and in the second part we will assess the use of inference to the best explanation to defend scientific realism.

7.1 Explanation

Explanation is supposed to tell us why things happen as they do. Many philosophers and scientists have thought that it is not enough for scientific theories to describe the world as it is, they should also tell us why it is that way. Consider the following explanations:

(i) The window is broken because it was hit by a stone.
(ii) The pressure of the gas rose because the volume was fixed and the temperature was increased.
(iii) They are not answering the phone because they want to get some work done.

The first is a straightforward causal explanation, the second appeals to the gas law that relates temperature, pressure and volume, and the third is a psychological explanation. Causal explanation attributes the structure of cause and effect to events to explain a phenomenon. Explanation in terms of laws, or *nomic* explanation, works by showing that the event had to happen given that the laws of nature are as they are. Psychological explanation works by appealing to our background knowledge of how people's behaviour is related to their desires and beliefs. Other specific kinds of explanation (sketches) are as follows:

(1) His violence is the result of a repressed Oedipal complex. (Psychoanalytical)
(2) Giraffes have long necks because it enables them to reach the leaves of tall trees for food, in others words it is an adaptation to the environment. (Evolutionary)
(3) The rise to power of the Nazis happened because of Germany's humiliation by the other European powers in the negotiations for the treaty at Versailles. (Historical)
(4) The handle of the spoon is curved to fit comfortably in the hand. (Functional)
(5) She is adaptable because she is a Pisces and that is a water sign. (Astrological)
(6) He died young because it is God's will. (Theological)
(7) The stone fell to Earth because its natural place is at the centre of the universe. (Teleological/Aristotelian)

(8) The rise of capitalism was an inevitable result of the instabilities in the organisation of production in eighteenth century Europe. (Marxist)

Philosophers and scientists have often been sharply divided over what kinds of explanation are legitimate. As we saw in Chapter 1, during the scientific revolution, explanations such as that given in (7) were widely rejected by natural philosophers, because they appeal to the natural place of things, or their proper function, without specifying the material causes of things. It is still widely thought that functional explanation is only legitimate when a plausible casual mechanism is available, even if only in outline. For example, in the case of the spoon (4), the underlying casual explanation is that people make it that way, and this becomes a combination of psychological and causal explanation. Similarly, talk of organisms having features that are adapted to the environment in evolutionary explanations is only thought to be legitimate because there is an underlying causal explanation. Namely that, given the pressure of a tough environment, and the fact that organisms pass on their basic physical characteristics via their genes, random variation in form among the individuals in a species results in different survival and successful reproduction rates. Those features organisms acquire that perform some function that increases their survival rate will also occur in a greater proportion of the organisms in the next generation; hence, over millennia, features such as legs, teeth and eyes are developed and so on. On the other hand, many have argued that the explanations offered by psychoanalysis, Marxism, astrology and other alleged pseudo-sciences are not genuine explanations because they offer no understanding of causal mechanisms that can be empirically tested. Whether or not psychological explanation is a species of causal explanation or of some other kind is also a controversial issue.

It might seem then that scientific explanation is usually or always causal explanation. Where there is an appeal to laws, this is to be supplemented with a causal account of why the laws hold. For example, the gas laws are to be explained in terms of the molecular motions that cause the phenomena of pressure and temperature. Some people do hold this view but, as we saw in Chapter 2, many philosophers worry about the notion of causation. The positivists, like Hume, objected to any idea of causal or necessary connections beyond observable regularities in the phenomena. In much of the modern science that so impressed them, explanations usually proceed from mathematical laws, such as Newton's laws of motion and gravitation or the gas law appealed to in (ii) above. They thought that Newton's talk of the gravitational *force* causing the Earth to be attracted to the Sun is only scientifically respectable because he gives a precise law governing its effect, which has testable consequences. Carnap and others criticised the theory of *vital forces* in biology, according to which living things harbour forces not found in the inanimate world, and which act counter to the general tendency of physical systems to move towards a random and unstructured state. They argued that it does not explain anything to postulate such forces because no detailed predictions can be derived from the laws that govern them.

Consequently, those who agree with Hume that what we call causation is just regularities in the phenomena think that causal explanation is reducible to nomic explanation. According to this view, in principle, the explanation in (i) above could be replaced by one in terms of the laws of mechanics – the laws of physics governing the behaviour of materials such as glass. They then argue that laws are nothing more than generalisations about how things behave. The idea is to combine Hume's regularity theory of causation with a regularity theory of laws, so doing away with metaphysical ideas of causal or nomic necessary connections between events.

7.1.1 *The covering law model of explanation*

It is often thought that scientists explain the phenomena we observe by discovering the laws of nature that govern these phenomena. One of the most influential theories of scientific explanation, which is due to Hempel, says that to give a scientific explanation of some phenomenon, event or fact is to show how it can be seen to follow from a law (or set of laws) together with a specification of initial conditions. So, for example, we can explain the darkening of the skies during a solar eclipse by using the laws of celestial mechanics, plus the relevant positions, masses and velocities of the Sun, the Earth and the other planets, to determine the path of the Moon during the relevant time period. Our background knowledge of the laws of optics then implies that the Moon will cast a shadow over a certain part of the Earth's surface when it passes between the Earth and the Sun. Of course, many scientific explanations are very complicated and a full specification of all the laws and initial conditions involved would be difficult if not impossible. However, many paradigmatic scientific explanations do seem to conform to this model, even if in practice an explanation sketch is all that is fully articulated. Hempel argued that genuine explanations, even in history or the social sciences, always involve reference to laws, even if they don't seem to be instances of the covering law model on the surface.

When talking about explanation, philosophers call whatever is being explained the *explanandum*; this might be a particular event, for example a solar eclipse, or a general class of phenomena, for example the planets having elliptical orbits or the scattering of light by prisms or water droplets. That which does the explaining is the *explanans*. So, for example, if the explanandum is the occurrence of high tides once a month, then the explanans would be the law of gravitation, and details of the mass of water in the oceans, and the mass and position of the Moon. Then we deduce from the law of gravitation that the Moon will exert a gravitational effect on the water on the Earth sufficient to cause high and low tides in various places depending on their position relative to the Moon. In the next section, I shall explain Hempel's first elaboration of the covering law model.

(1)   The deductive-nomological model

According to this model of explanation (the DN· model), the explanandum must be *deduced* from a law or laws of nature, plus *background facts* and *initial conditions*. (The term 'nomological' means 'pertaining to laws of nature'.)

Schematically:

laws	$L_1, L_2, \ldots, L_m$
initial conditions	$C_1, C_2, \ldots, C_n$
entail	
explanandum	$O_1, O_2, \ldots, O_p$

where m, n and p are natural numbers, and there are m laws, n initial conditions and the explanandum is a series of p observations.

The DN model imposes the following *logical conditions* on explanations:

(i)   The explanans must deductively entail the explanandum.
(ii) The deduction must make essential use of general laws.
(iii) The explanans must have empirical content.

(i) says that putative explanations are effectively deductively valid arguments; the explanandum must follow deductively from the explanans. (ii) says that to be an explanation an argument must include one or more laws among its premises in such a way that the argument would not be valid without them; this is to ensure that a pseudo-scientific explanation that incorporates laws in an inessential way to give the appearance of scientific explanation will not satisfy the DN model. (iii) says that the explanans, that is the laws and the other premises concerning initial and background conditions, must be empirically testable. Finally, we must add the following *empirical condition*:

(iv) The sentences in the explanans must be true.

(iv) ensures that the argument is sound because it is seems obviously unsatisfactory to explain something by appealing to a false pro-position. For example, we can't explain the fact that a particular chemical compound dissolves in water by deducing it from the law that states that all compounds dissolve in water, because that law is false. Of course, knowledge of which laws of nature are true is fall-ible. If it turns out that one of our cherished laws is false, then accord-ing to the DN model we thought we had an explanation but in fact we didn't.

(2)   Problems with the covering law account

Most of the objections to the covering law model of explanation are designed to show that Hempel's conditions are not *sufficient*; in other words, that an argument may satisfy all of them but still not count as a scientific explanation. Some people also argue that the conditions are not even *necessary*, in other words that a proper scientific explanation need not satisfy them all. Some of the following objections overlap with each other but it is useful to treat them separately.

(2.1) Irrelevance

This is where we have an argument that satisfies the covering law model but where, intuitively, part of the explanans is not a relevant explanatory factor. For example, we might offer the following explanation:

> All metals conduct electricity
> Whatever conducts electricity is subject to gravity
> Therefore, all metals are subject to gravity

This argument is sound, and the premises are general laws, but the fact that metals conduct electricity is irrelevant to their being subject to gravity. It is easy enough to generate similar examples.

> All salt dissolves in holy water
> A sample of salt was placed in some holy water
> Therefore, the sample of salt dissolved

Again, that the water was holy does not explain why the salt dissolved. The problem of irrelevance may be countered by adding a condition to the effect that the premises of proper explanations must be relevant and have no gratuitous extra clauses or laws. However, explicating the notion of relevance needed is not a trivial matter.

(2.2) Pre-emption

Pre-emption is when an event that was going to happen for some reason happens earlier for a different reason. Consider the following:

> (a) Everyone who eats a pound of arsenic dies within 24 hours
> (b) Marge ate a pound of arsenic
> Therefore, Marge died within 24 hours

(a) certainly seems to be a law, but suppose, although (b) is true and Marge did indeed die within 24 hours, she was in fact run over by a bus. Although all the conditions of the DN model are satisfied, the law does not explain how she died.

(2.3) Overdetermination

An event is said to be overdetermined when more than one set of causal conditions are in place but each of them is sufficient to bring it about. For example, suppose someone is shot in the head at the same time as being electrocuted; then their death is overdetermined. Now, consider the following:

> All people who do not have sex do not become pregnant
> Nick (who is a man) did not have sex
> Therefore, Nick did not become pregnant

Obviously Nick not being with child is overdetermined by his not having sex and being a man. His not having sex is not what explains his failure to conceive. However, this argument satisfies the conditions of the DN model.

(2.4) Symmetry

Many scientific laws are laws of co-existence, in other words they limit what possibilities can be realised simultaneously. The gas laws are of this form, because they constrain the values of the pressure, volume and temperature of a gas at a given time. However, when we have such a law we seem to run into trouble because we can generate cases where two events seem to explain each other. For example, suppose it is a law that all animals with hearts also have livers and all animals with livers also have hearts. Then from the observation that some particular animal has a heart we can explain why it has a liver. However, we could equally observe that it has a liver and use this fact with the law above to explain why it has a heart. Intuitively, neither of these explanations is satisfactory. Now consider the following:

> A gas is sealed in a container of fixed volume and heated strongly. If the volume of a gas is kept constant then its temperature is directly proportional to its pressure.
> Therefore, the pressure of the gas rose.

This seems to be an adequate explanation yet we could just as easily reverse the explanatory order while still satisfying the DN model:

A gas is sealed in a container of fixed volume and its pressure rises.
If the volume of a gas is kept constant then its temperature is directly proportional to its pressure.
Therefore, the temperature of the gas rose.

However, this second explanation is intuitively wrong because the temperature increase caused the pressure rise and not the other way around.

(2.5) Prediction and explanation

Hempel advocates 'the thesis of structural identity' according to which explanations and predictions have exactly the same structure; they are arguments where the premises state laws of nature and initial conditions. The only difference between them is that, in the case of an explanation we already know that the conclusion of the argument is true, whereas in the case of a prediction the conclusion is unknown. For example, Newtonian physics was used to predict the return of Halley's comet in December of 1758, and once this was observed the same argument explains why it returned when it did.

However, there are many cases where the observation of one phenomenon allows us to predict the observation of another phenomenon without the former explaining the latter. For example, the fall of the needle on a barometer allows us to predict that there will be a storm but does not explain it. Similarly, the length of a shadow allows us to predict the height of the building that cast it, and if we know the period of oscillation of a pendulum we can calculate its length, but in both these cases the latter explains the former and not the other way round. The cases of pre-emption and overdetermination above are also examples where, contrary to the symmetry thesis, there seem to be adequate predictions that fail to be adequate explanations. Furthermore, there seem to be adequate explanations that could not be predictions. For example, evolutionary theory explains, but it cannot usually make specific predictions because evolutionary change is subject to random variations in environmental conditions and the morphology of organisms. Probabilistic explanations offer further examples where prediction and explanation seem to come apart since, when the probability conferred by the explanans on the explanandum is low, we cannot predict that the explanandum is even likely to happen, although we can explain why it did afterwards.

(3) The inductive statistical model

The DN model of explanation is of no use when it comes to using statistical or probabilistic laws to explain things. In the social sciences, probabilistic laws may be all that is available. For example, it seems to be a law of human societies that a low per capita income is correlated with a high rate of infant mortality. Hempel's Inductive-Statistical (IS) model assumed that in cases of probabilistic explanation, the law plus auxiliary conditions makes the explanandum very likely:

Law	prob (O/F) is very high
Conditions	F_m
Make likely	=============
Explanandum	O_n

However, a number of counterexamples have been found to this model, where a probabilistic cause of some phenomenon only gives it a low probability of occurring. For example, there is a disease called paresis, which is only contracted by those already suffering from syphilis, but the probability of getting it is still low among syphilis sufferers. Therefore, although it is some explanation of someone's getting paresis that they already have syphilis, this explanation will not satisfy the IS model, and so the latter cannot be a necessary constraint on probabilistic explanation. (Notice that there is a tradition in the literature about explanation to make use of morbid and/or medical examples wherever possible.) There are also cases where events are highly probabilistically correlated but where neither is part of the explanation of the other, which shows that the IS model is not sufficient either. For example, the fall of a barometer needle is highly correlated with the occurrence of rain but neither event explains the other, rather they are both explained by the presence of a low pressure system (which is their 'common cause').

7.1.2   *Other theories of explanation*

In an attempt to avoid the problems faced by the covering law account, some philosophers, such as Wesley Salmon and David Ruben, have adopted a causal theory of explanation, according to which to explain something is to give a specification of its causes. According to this view, explanations are not arguments and they need not involve laws. This seems to avoid many if not all the problems above; for example, the holiness of some water plays no causal role in the boiling of it, hence even though it may be a true generalisation that all holy water boils at 100° Celsius, to deduce that some particular holy water will do so from this fact does not really explain why it does so, because the right causes have not been specified. On the other hand, if causes are not reducible to laws and hence to regularities, what are they? Furthermore, there are cases where one law is used to explain another with no mention of causation, as with, for example, the explanation of Kepler's laws with Newton's law of gravitation, which is silent as to what causes gravitational phenomena.

The defender of the covering-law account of explanation faces another challenge, which is to characterise precisely what a law of nature is without invoking anything metaphysical. Laws are supposed to be simply universal generalisations that are true of the actual universe, but not all such generalisations are counted as laws. The problem is to explain the difference between a law and an accidentally true generalisation. For example, consider the following claims:

(1) All solid spheres of gold have a diameter of less than 100 miles.
(2) All solid spheres of enriched plutonium have a diameter of less than 100 miles.

It is plausible that these are both true universal generalisations about the actual universe. However, (1) is accidentally true in the sense that it could have been false. There is no particular reason to think that it is impossible for there to be a solid sphere of gold with a diameter of 100 miles. On the other hand, (2) is true because it is a law of nature that any amount of enriched plutonium remotely approaching the size of such a sphere would immediately undergo a chain reaction resulting in its immediate and catastrophic decay.

Counterfactual conditionals are statements such as 'if you had dropped the glass it would have broken', where the antecedent is false. Laws of nature seem to support counterfactuals, as do causal claims, but accidental universal generalisations do not. For example, the gas law supports the counterfactual 'if the gas had been heated and its volume kept constant its pressure would have risen'; and the claim that hitting fragile objects causes them to break implies that 'if the ball had hit the window it would have smashed'. However, the true generalisation 'all the coins in my pocket are silver (in colour)', does not imply the truth of 'if this copper coin had been in my pocket it would have been silver'. There is a link between the fact that laws support counterfactuals and their use in explanations. After all, consider a particular coin in my pocket. We can deduce from the generalisation above that it is silver, but this does not explain why it has that colour. Some philosophers (such as David Armstrong, see the Further reading section at the end of this chapter) have addressed this problem by abandoning the regularity theory of laws and arguing that laws must be understood as necessary connections of some kind.

An alternative account of explanation to one invoking a metaphysics of causes or lawlike necessary connections is advocated by van Fraassen. He points out that there is always a pragmatic component to explanation in so far as explanations are addressed to people's interests. For example, suppose someone asks 'why did the dog bury the bone?'. This might mean 'why did the dog, not some other animal, bury the bone?, 'why did the dog bury, rather than eat, the bone?', or perhaps 'why did the dog bury the bone, but not a ball?'. The point is that what counts as an explanation in a given context depends upon the possible contrasts that the questioner has in mind. Hence, van Fraassen argues that the explanatory power of a theory or hypothesis is dependent on the context of a why-question. As we shall see, it is because he thinks that explanation is a pragmatic feature of theories that he rejects the link between explanation and inference.

7.2   Inference to the best explanation

IBE is supposed to be a rule of inference according to which, where we have a range of competing hypotheses, and all of which are empirically adequate to the phenomena in some domain, we should infer the truth of the hypothesis that gives us the best explanation of those phenomena. Gilbert Harman introduced the term 'Inference to the best explanation' in an article of that name in the *Philosophical Review* in 1965, but it is sometimes also known as 'abduction' – following the terminology of Charles Pierce (1839–1914). It certainly seems that, in everyday life when faced with a range of hypotheses that all account for some phenomenon, we usually adopt the one that best explains it. For example, you ring your friend's doorbell and there is no answer. The following hypotheses all predict this:

(1) Your friend has become paranoid and thinks that enemy agents are ringing the bell.
(2) Your friend has suddenly gone deaf.
(3) Your friend has been pretending to live there, but in fact lives somewhere else.
(4) Your friend is out.

Normally we would infer (4) was correct because it offers a simple explanation of the data that coheres with our other beliefs. Advocates of inference to the best explanation argue that much of our inductive (that is, non-deductive) reasoning in everyday life works this way.

Here is an example from van Fraassen: suppose I hear scratching in the wall of my house, the patter of little feet at midnight, and cheese keeps disappearing. Wouldn't I infer that a mouse has come to live with me even if I never see it? (van Fraassen 1980: 19). This inference has the structure, if p then q, q therefore p, in other words, we know that if there is a mouse then there will be droppings, noises and other observable evidence, and we observe the evidence and so infer the existence of a mouse. However, consider the following: if something is a cat, then it is a mammal, Bess is a mammal therefore Bess is a cat; this is deductively invalid because it is possible for the conclusion to be false when the premises are both true, for example, if Bess is a dog (this is the fallacy called *affirming the consequent*). Similarly, there is no contradiction in supposing that there is no mouse in my house despite the evidence, so that inference to the best explanation is also deductively invalid.

Nevertheless, it is hard to see how we could get by without such inferences, and in science there are many examples where IBE seems to be used to choose between theories. As mentioned above, evolutionary theory is much better at explanation than prediction, as are specific theories about unobserved but observable entities like dinosaurs. When it comes to theories of the origins of stars or the nature of the inside of the Earth, how else could we decide which to adopt other than by virtue of their relative explanatory power? This is, of course, too brief without an account of what makes a good explanation, and how a host of different features contribute to the attribution of explanatory success to a theory, including some of the following:

(1) Otherwise surprising phenomena are to be expected if the hypothesis is true.
(2) Predictions of empirical consequences must be inferred from the hypothesis and tested and confirmed.
(3) Simple and natural hypotheses are to be favoured.
(4) Hypotheses that cohere with metaphysical views are to be favoured.
(5) Unifying power and wideness of scope of the hypothesis are to be favoured.
(6) Hypotheses that cohere with other scientific theories are to be favoured.

Now, there are two ways in which inference to the best explanation is used to argue for scientific realism, and I shall call them *local* and *global*.

7.2.1 *The local defence of scientific realism*

As I pointed out in the previous chapter, realists often argue that, in scientific practice, underdetermination is broken because empirical equivalence does not imply evidential equivalence. For example, consider the theory of molecular structure, according to which chemical compounds, such as water and sulphuric acid, are made of characteristic combinations of atoms, in this case water is two hydrogen and one oxygen, and sulphuric acid is two hydrogen, one sulphur and four oxygen atoms. This theory includes the details of the way electrons are arranged in orbits and the number of electrons in the outermost orbit of an atom determines the way it can bond with other atoms. The theory unifies a vast array of experimentally determined facts about chemical phenomena, such as the energy it takes to break down particular compounds, and why certain elements only combine in certain proportions. Realists argue that we accept this theory, not only because it predicts the phenomena we observe, but also because it explains them; there really are tiny atoms, with nuclei and electrons orbiting them, they obey certain physical laws and this is why the behaviour of chemical substances is as it is. Another theory, which just says that the world is as if there are such atoms, predicts all the same phenomena but does not explain them.

It seems that many scientists think that to regard a theory as the most explanatory successful is therefore to have a good reason for choosing that theory over its rivals. Hence, the argument goes, if we accept the rationality of scientific practice then we have to accept the rationality of IBE. If the theory in question refers to unobservable entities then accepting its truth entails accepting the existence of these entities, therefore the practice of IBE in science commits us to realism. It will be helpful in what follows to have in mind some examples. In each case I shall state the phenomena to be explained and the hypothesis that explains them:

(1) A trail of vapour in the sky, the sound of a jet engine, blips on a radar screen.
    There is a plane in the sky at high altitude that cannot be seen.
(2) Large fossilised bones from no known living animals, huge footprints from no known living animals.
    Dinosaurs walked the Earth.
(3) Astronomical data and observations consistent with the existence of moons of Jupiter.
    Jupiter has moons.

(4) Tracks in a cloud chamber, dots on a television screen, electrical phenomena.
 There are electrons.
(5) Reports of alien abduction, UFO sightings and so on.
 There are aliens.

In (1) the hypothesis concerns the existence of an entity that it is practically possible to observe – one could fly after it in another jet and see it, or track it on the radar until it lands and look at it then. This example is clearly analogous to that of the mouse in the wainscoting. In (2) the hypothesis concerns entities that we cannot observe because of our relative location in time; however, we could observe them if we were able to travel (backwards) in time. In (3) the hypothesis concerns entities that we cannot observe because of our relative location in space; however, we could observe them if we were able to travel in space (more effectively). In (4) the hypothesis concerns entities that we cannot observe because of our constitution as observers; that is, because of the anatomy and physiology of our perception. In (5) the hypothesis concerns entities that many argue have not actually been observed, despite delusional claims to the contrary, although they are presumably observable if they exist.

For the moment ignore (5). The difference between van Fraassen and the scientific realist is that van Fraassen accepts the existence of the entities in (1)–(3), but not in (4), while the realist accepts the existence of all of them. Even though we have never observed dinosaurs or the moons of Jupiter, and even if it happens to be the case that we never will, van Fraassen is committed to the view that, since they are both observable, then commitment to the theories that describe them and belief in their empirical adequacy entails the belief that they exist. Now it has seemed to many of his critics that van Fraassen thinks that one should infer the truth of the explanans in (1)–(3) but not in (4) because, in the latter case, empirical adequacy is the not the same thing as truth; since, in principle, we cannot observe electrons, there could be some other explanation of the phenomena that is actually true. For example, according to Stathis Psillos (1999, Chapter 9), van Fraassen attempts to show that IBE cannot provide an epistemic warrant for hypotheses about unobservables, whereas it can for hypotheses concerning only observables. Psillos understands van Fraassen as advocating a rule of 'inference to the empirical adequacy of the best explanation'. To return to one of the examples discussed above, we might think that van Fraassen is saying that believing that all phenomena are as if a plane is overhead is equivalent to believing that there really is a plane overhead, but that there is a gap between the corresponding beliefs about electrons, say, since they are unobservable. This is to say that the reason that IBE is respectable in the context of observable entities is that, in such cases, an inference to the empirical adequacy of a hypothesis is equivalent to an inference to its truth. However, in fact, van Fraassen's attack on IBE does not discriminate between explanations that do and do not postulate unobservables, and he denies that IBE is compelling even in the case of the observable. Before we consider this we need to see how realists have sought to defend scientific realism with IBE in a different way.

7.2.2 The global defence of realism

IBE is also used by scientific realists to defend realism at the global level, where the explanandum is the success of science as a whole. This is the so-called ultimate argument for scientific realism, also known as the 'no-miracles argument', which was famously presented by Hilary Putnam: 'the positive argument for realism is that it is the only philosophy that doesn't make the success of science a miracle' (Putnam 1975a: 73). In particular, it is the success of science in predicting new and surprising phenomena, and in the application to technology that realists argue would be miraculous if the theories were not, in general, correctly identifying the unobservable entities and processes that underlie what we observe.

So the idea is that the overall predictive and instrumental success of science is inexplicable by anything but a realist view, so strictly speaking this is an inference to the *only* explanation. A similar form of argument is to be found in Jack Smart's *Philosophy and Scientific Realism*: 'If the phenomenalist about theoretical entities is correct we must believe in a *cosmic coincidence*' (Smart 1963: 39). The coincidence in question would be that instruments and devices such as electron microscopes and microwave ovens mysteriously behave just like they would if there were atoms and electromagnetic waves. Isn't it more plausible to suppose that there really are unobservable entities as our theories say there are?

Notice the naturalism that is a feature of this defence of scientific realism. Many scientific realists argue that there is no fundamental difference between evaluating scientific theories and evaluating philosophical theories about science: 'philosophy is itself a sort of empirical science' (Boyd 1984: 65). Van Fraassen summarises the attitude of many realists towards the debate about scientific realism: 'If we are to follow the same patterns of inference with respect to this issue as we do in science itself, we shall find ourselves irrational unless we assert the truth of the scientific theories we accept' (van Fraassen 1980: 19). Hence, scientific realism is seen as a scientific hypothesis that is supposed to explain empirical facts about the history of science itself.

The global defence of realism is made more sophisticated by citing specific features of scientific methodology and practice and arguing that they are particularly in demand of explanation, and furthermore that realism offers the best or only explanation. Richard Boyd (1985, for example) has argued that, in particular, we need to explain the overall instrumental success of scientific methods across the history of science. All parties in the realism debate agree that:

(i) Patterns in data are projectable from the observed to the unobserved using scientific knowledge, which is to say that induction based on scientific theories is reliable.

(ii) The degree of confirmation of a scientific theory is heavily theory-dependent, in the sense that background theories inform judgements about the extent to which different theories are supported by the available evidence.

(iii) Scientific methods are instrumentally reliable, in other words, they are reliable ways of achieving practical goals like prediction and the construction of technological devices.

Boyd and other realists go on to argue that these features of science would be utterly mysterious if the theories involved were not true or approximately true. For example, consider the theory in biology according to which human cells have a complex structure including a nucleus and a semi-permeable membrane at the cell wall to allow the passage of proteins and nutrients. This theory is confirmed by techniques involving optical and electron microscopes, yet those instruments are constructed according to the laws of optics and quantum mechanics respectively. The only explanation of the reliability of the latter as background theories is that they correctly describe the way light and electrons respectively behave. (It is ironic that it makes the theory-ladenness of scientific methodology and the confirmation of theories a reason for adopting realism, when it is usually regarded as telling against realism, cf. Chapter 4).

Another feature of scientific practice that realists have long argued cannot be explained by antirealists is the persistent and often successful search for unified theories of diverse phenomena. The well-known 'conjunction objection' against antirealism is as follows: Consider two scientific theories, T and T', from different domains of science, say chemistry and physics. That T and T' are both empirically adequate does not imply that their conjunction $T \& T'$ is empirically adequate; however, if T and T' are both true this does imply that $T \& T'$ is true. So, the argument goes, only realists are motivated to believe the new empirical consequences obtained by conjoining accepted theories. However, it is claimed that, in the course of the history of science, the practice of theory conjunction is widespread and a reliable part of scientific methodology. Therefore, if scientists are not irrational, since only realism can explain this feature of scientific practice, then realism must be true.

Van Fraassen's (1980: 83–87) response to this is simply to deny that scientists ever do conjoin theories is this manner; rather, he argues, the process of unifying theories is more a matter of 'correction' than 'conjunction'. Furthermore, he argues that scientists have pragmatic grounds for investigating the conjunction of accepted theories in the search for empirical adequacy. It is certainly true that the above simplistic view of theory conjunction does not do justice to the complexity of the practice of conjoining real theories. In some cases, the conjunction of two theories will not even be well-formed because they adopt such different approaches, as with general relativity and quantum mechanics, for example. Note also that if we suppose that T and T' are both approximately true, this does not imply that their conjunction $T \& T'$ is also approximately true. For example, suppose T is Kepler's laws of planetary motion, and T' is Newtonian mechanics; then, because the former says that the planets move in perfect ellipses, whereas the latter says that their motions are more complex, T & T' is actually inconsistent and therefore cannot even be approximately true. (For more on approximate truth see 8.1.)

There is a general argument to show that constructive empiricism can account for the worth of any aspect of the practice of science that realism can. Suppose some feature of scientific practice is claimed by the realist to have, as a matter of fact, produced instrumental success, and the realist claims to have an explanation or justification of that feature. The antirealist can simply point out that the history of science provides inductive grounds for believing in the pragmatic value of that feature of scientific practice. Similarly, van Fraassen offers an account of the pragmatics of science that attempts to account for (i), (ii) and (iii) above with the fact that the background theories are empirically adequate.

Furthermore, van Fraassen objects that the realist's demand for explanation presupposes that a lucky accident or coincidence can have no explanation at all, whereas he thinks that coincidences may have explanations in a certain sense (van Fraassen 1980: 25). His example is meeting his friend in the market:

> It was by coincidence that I met my friend in the market – but I can explain why I was there, and he can explain why he came, so together, we can explain how this meeting happened. We call it a coincidence, not because the occurrence was inexplicable, but because we did not severally go to the market in order to meet. There cannot be a requirement upon science to provide a theoretical elimination of coincidences, or accidental correlations in general, for that does not even make sense.
>
> (van Fraassen 1980: 25)

However, this seems to miss the point of the no-miracles argument; the realist's claim is just that explanation of the repeated predictive success of scientific theories in terms of coincidence or miraculous luck is an unacceptable and arbitrary explanation, especially given the availability of the realist's alternative. Similarly, if I kept meeting my friend in the market unexpectedly and some explanation other than mere coincidence was available, then I might be inclined to adopt it.

Finally, van Fraassen offers his own Darwinian explanation for the existence of predictively successful scientific theories:

> [T]he success of current scientific theories is no miracle. It is not even surprising to the scientific (Darwinist) mind. For any scientific theory is born into a life of fierce competition, a jungle red in tooth and claw. Only the successful theories survive – the ones which *in fact* latched on to actual regularities in nature.
>
> (van Fraassen 1980: 40)

As a result, realists have retrenched to the claim that realism, although not the only, is at least the best explanation of the success of science. They go on to point out that van Fraassen's explanation is a *phenotypic* one; it offers a selection mechanism for how a particular phenotype (empirically successful theories) has become dominant in the population of theories. However, this does not preclude a *genotypic* explanation of the underlying features that make some theories successful in the first place: 'a selection mechanism may explain why all the selected objects have a certain feature without explaining why each of them does' (Lipton 1991: 170). For example, we can explain in two ways why a particular giraffe has a long neck: we can point out that giraffes that had short necks did not survive as well, or we can explain how its genes and constitution give it a long neck. These explanations are compatible. Hence, realists accept van Fraassen's phenotypic explanation but they also accept that theories are approximately true as a genotypic explanation of their instrumental reliability. Hence, Peter Lipton (1991: 170ff) argues that the latter explains two things that van Fraassen's explanation does not: (a) why a particular theory that was selected is one having true consequences; and (b) why theories selected on empirical grounds go on to have more predictive successes.

Antirealists have another response to this argument available to them, which is just to argue that the predictive success of theories is explained by their empirical adequacy. The realist argues that the empirical adequacy of theories is itself in need of explanation in terms of the truth of those theories, but then why do we not need to explain further the truth of theories, perhaps in terms of God having willed it? Leplin argues that the truth of a theory does not itself need an explanation because it is either explained by another deeper theory or 'is the way of the world' (Leplin 1997: 33). However, surely either of these possibilities is available to explain the empirical adequacy of a theory that may be explained by the empirical adequacy of a deeper theory, in the same way as the approximate empirical adequacy of Newtonian mechanics is explained by that of relativity theory, or it may just be the way of the world. On the other hand, perhaps it is better for van Fraassen simply to deny that the predictive success of particular theories needs any explanation at all, and rely upon his arguments against IBE, to which we will turn in the next section. In any case, a more fundamental criticism of the use of IBE at the global level was made by Laudan (1981: 133–135) and Arthur Fine (1984: 85–86), both of whom pointed out that since it is IBE involving unobservables that is in question in the realism debate, it is circular to appeal to the explanatory power of scientific realism at the meta-level to account for the overall success of science because realism is itself a hypothesis involving unobservables. Hence, it is argued that the global defence is question begging.

There is a similarity here with the inductive vindication of induction that we considered in 2.2 (8). Richard Braithwaite (1953: 274–278), and Carnap (1952), defended the view that the inductive defence of induction – induction has worked up to now so it will work in the future – was circular but not viciously so, because it is *rule circular* not *premise circular*. In the case of IBE such a view has been defended recently by David Papineau (1993, Chapter 5) and Psillos (1999, Chapter 4). The idea is that premise circularity of an argument is vicious because the conclusion is taken as one of the premises; on the other hand rule circularity is when the conclusion of an argument states that a particular rule is reliable, but that conclusion only follows from the premises when that very rule is used. Now notice that the global defence of realism is rule- but not premise-circular. The conclusion that the use of IBE in science is reliable is not a premise of this defence of realism, but the use of IBE is required to reach this conclusion from the premises that IBE is part of scientific methodology and that scientific methodology is instrumentally reliable.

It is conceded that, although it is not viciously circular, this style of argument will not persuade someone who totally rejects IBE. However, what the argument is meant to show is that someone who does make abductive inferences can show the reliability of their own methods. So, it seems that IBE is on a par with inductive reasoning; it cannot be defended by a non-circular argument, but recall that even deduction cannot be defended by a non-circular argument either (see 2.2 (8)). Hence, realists may claim that, although they cannot force the non-realist to accept IBE, they can show that its use is consistent and then argue that it forms part of a comprehensive and adequate philosophy of science. As we shall see, van Fraassen has advanced arguments to undermine even this claim. If his arguments work then the realist cannot appeal to IBE to defend realism either at the local or the global level.

7.2.3 Van Fraassen's critique of inference to the best explanation

Van Fraassen offers several arguments against the idea that IBE is a rule of inference. Here are two.

(1) The argument from indifference

The argument from indifference is roughly that, since there are many ontologically incompatible yet empirically equivalent theories, it is highly improbable that the true theory is in the class that we have to choose between; hence, it is highly improbable that the best explanatory theory is true. This argument appeals to the existence of empirical equivalents to any theory that we have. In the discussion of the underdetermination problem in Chapter 6 it was concluded that the constructive empiricist may also be threatened by the existence of empirical equivalents. Similarly, Psillos (1996) argues that the argument from indifference works as much against constructive empiricism as realism, since any finite set of theories that we consider is just as unlikely to contain an empirically adequate theory. However, this does not help defend IBE.

(2)   The argument from the best of a bad lot

This argument is that some 'principle of privilege' is required if we are to think that the collection of hypotheses that we have under consideration will include the true theory. The best explanatory hypothesis we have may just be the best of a bad lot, all of which are false. In other words, this argument challenges the proponent of IBE to show how we can know that none of the other possible explanations we have not considered is as good as the best that we have. Unless we know that we have included the best explanation in our set of rival hypotheses, even if it were the case that the best explanation is true, this would not make IBE an acceptable rule of inference.

Realists tend to bite this bullet and argue that scientists do have privilege, which issues from background knowledge. Theory choice is informed by background theories, which narrow the range of hypotheses under consideration, and then explanatory considerations help select the best hypothesis. Furthermore, they argue that both the realist and the constructive empiricist need privilege, because the constructive empiricist needs to assume that the empirically adequate theory is among the ones considered in order to have warranted belief in the empirical adequacy of the chosen theory. Hence the dispute can only be about the extent of that privilege.

7.2.4   Selective scepticism?

[T]he primary issue in the defence of Scientific Realism is selective scepticism: epistemic discrimination against unobservables; unobservable rights.

(Devitt 1991: 147)

Suppose we grant that van Fraassen's arguments against IBE are compelling. Realists argue that any antirealist who is not an outright sceptic will need to use some criteria to distinguish their preferred theories from those that are empirically equivalent to them. The problem for van Fraassen is that he seems to be left with no grounds even for believing a theory to be empirically adequate. In the previous chapter, I distinguished between strong and weak underdetermination. If we were entirely restricted to empirical features of theories then we would be unable to choose among theories that are merely weakly empirically equivalent. For example, recall that for any theory T, we can define the theory $T\#$ which says that everything that is observed before some arbitrary point in time is in accordance with T, but not afterwards.

The point is that there are indefinitely many empirically distinct theories that agree about what has been observed *so far*. What is the warrant for the antirealist to infer that a particular theory is even empirically adequate? Any considerations that allow the antirealist to infer that a theory is empirically adequate must be non-empirical. So realists can argue that either: (i) van Fraassen has no resources to break the underdetermination of which theory is empirically adequate by the available evidence, or even to warrant ordinary inductive inferences or belief in observable entities, and so he must be an out and out sceptic; or (ii) he must adopt an arbitrary scepticism about IBE for unobservables while endorsing IBE for theories about observables and for inferences to the empirical adequacy of theories about unobservables (in order to warrant belief in observable entities and in judgements of empirical adequacy in line with constructive empiricism).

Now I think it is clear that there can be an extra problem with IBE over and above the problem of induction. When we routinely use IBE to go beyond the observed phenomena, we do so without introducing new ontological commitments. In the case of van Fraassen's example, *we already believe that mice exist*, that is we use IBE to conclude new facts about tokens of types that are already included within our onto-logical commitments. It may be objected that the particular mouse in question is not part of our ontological commitments; however, to admit the existence of a new type of entity is what is at stake in the realism debate and this goes beyond what is at stake in the everyday use of IBE. The legitimacy of introducing new ontological commit-ments with IBE is not uncontroversial even in the case of observable entities. Recall (5) of 7.2.1 above: a variety of phenomena occur that are empirical consequences of the hypothesis that there are aliens landing on Earth and kidnapping and experimenting with human beings against their will. Does this compel us to believe in the exist-ence of aliens? Certainly not, for there are other (weakly empirically equivalent) hypotheses that also explain the data that we may choose to accept instead. One reason for doing so may be that these alternatives do not require us to believe in some new type of entity. The constructive empiricist may add that the only evidence that would really convince most of us of the existence of aliens would be a direct sighting, personally or by people that we trust; but then in the domain of the unobservable, direct sightings cannot occur and the strong underdetermination problem is not just a matter of lack of evidence. If we already accept the existence of two species of aliens, then inferring the existence of a third would be much less contentious. Of course, it may seem inappropriate to compare aliens with electrons since the former are supposed to be observable; hence we demand that we should observe them ourselves because it is possible to do so. However, the point of the example is that, even in the case of observable entities, we have to be very careful about accepting the existence of a new type of entity.

In any case, realists argue that van Fraassen needs to use a rule of ampliative inference to have warranted belief in the empirical adequacy of a theory. Otherwise what grounds has he for making the particular inferences that he does and abstaining from others? There is an issue that has been buried throughout my discussion so far that I now want to raise. When we are assessing inferences like (1)–(5) in 7.2.1 above, are we asking (a) whether it is rational to believe in the entity/explanation in question, or (b) whether it is irrational not to believe in the entity/explanation in question? Realists often seem to think that, given that a particular explanation is agreed to be the best explanation of the phenomena in question and supposing its adequacy as an explanation, it is irrational not to adopt it. On the other hand, van Fraassen presents constructive empiricism not as a doctrine that must be adopted on pain of irrationality, but as a position that may be adopted while accounting for all that we need to about science.

Van Fraassen elaborates this in terms of his 'New epistemology'. In his book *Laws and Symmetry*, he makes it clear that he regards rationality as a *permission* term, not an *obligation* term (van Fraaasen 1989: see, for example, 171–172). He cites the distinction between so-called Prussian and English law. Apparently, the former forbids that which is not specifically allowed, while the latter allows anything that is not specifically forbidden. There are analogously two conceptions of rationality. In the Prussian model: 'what it is rational to believe is exactly what one is rationally compelled to believe'. In the English model: 'rationality is only bridled irrationality [. . .] what it is rational to believe includes anything that one is not rationally compelled to disbelieve' (van Fraassen 1989: 171–172). Van Fraassen opts for the latter view, which is called *voluntarism*. Indeed, according to van Fraassen, IBE may be indispensable in acquiring reasonable expectations and thus may be *pragmatically* indispensable. However, since what it is reasonable to believe will depend on pragmatic factors, this is not the same as endorsing its status as a rule of reasoning that issues in rationally compelled belief. His attack is thus against the realist who claims that IBE 'leads to truth' (van Fraassen 1989: 142–143), and not against IBE itself (or belief in the possibility that it might lead to truth), as many realists assume. Hence, van Fraassen's argument is directed against IBE understood as a *rule* of inference, not as an inferential practice:

> Someone who comes to hold a belief because he found it explanatory, is not *thereby* irrational. He becomes irrational, however, if he adopts it as a rule to do so, and even more if he regards us as rationally compelled by it.
>
> (van Fraassen 1989: 132)

Where does this leave us? For the sake of argument, let us grant van Fraassen the following: (a) constructive empiricism is prima facie coherent as a view of science and avoids the immediate problems of positivism by accepting a literal construal of theoretical vocabulary; (b) the vagueness of the observable/unobservable distinction does not alone obviate its epistemic significance; (c) granting epistemic significance to the observable/unobservable distinction that the observed/ unobserved distinction does not have is not incoherent; (d) the theory-laden nature of our discourse about the world, and the use of theoretical science to describe what is observable does not alone commit us to scientific realism; (e) IBE is not a compulsory rule of inference, hence scientific realism is not compelled by the canons of rational inference; and (f) voluntarism means that ampliative inferences are not irrational and hence, constructive empiricism need not collapse into total scepticism.

All we have shown so far is that constructive empiricism is a possible view of science and that IBE does not compel us to be realists. Constructive empiricism may be a consistent alternative position to scientific realism, and it may even be compatible with the practice of science. However, the mere availability of constructive empiricism will not persuade many realists to accept it and abandon scientific realism, especially given that van Fraassen concedes that scientific realism is not irrational. After all, for many philosophers, scientific realism is the natural position that is held prephilosophically and not primarily because of the persuasiveness of the arguments in its favour. Suppose that van Fraassen's voluntarism allows him to believe certain propositions and abstain from believing others, merely on the grounds that the latter are about unobservables. The realist will still object that to accept the risks involved in believing more than is logically implied by the data, but not to accept abduction to the existence of unobservables, is arbitrary and capricious. Hence, argues the realist, the only reason that van Fraassen does not advocate either total scepticism, or for that matter scientific realism, is mere prejudice.

However, van Fraassen argues that if we *need* go no further than belief in the empirical adequacy of theories to account for the nature and practice of science, then, if we do go further, we take an unnecessary epistemic risk for no extra empirical gain. Hence, his infamous slogan: 'it is not an epistemological principle that one might as well hang for a sheep as for a lamb' (van Fraassen 1980: 72). On the other hand, the realist insists that realism offers benefits that constructive empiricism does not. After all, the realist has explanations to offer for the phenomena we see around us and may claim, as Psillos does, that science has 'push[ed] back the frontiers of ignorance' (Psillos 1996: 42), whereas the constructive empiricist cannot.

Van Fraassen concedes that:

A person may believe that a certain theory is true and explain that he does so, for instance, because it is the best explanation he has of the facts or because it gives him the most satisfying world picture. That does not make him irrational, but I take it to be part of empiricism to disdain such reasons.

(van Fraassen 1985: 252)

Hence, van Fraassen is content to argue that *empiricists* should not be scientific realists and should adopt constructive empiricism because, from an *empirical* point of view, the extra strength of the realist position is illusory. Where we are dealing with the unobservable there is no further confrontation with experience that may tell in favour of the truth of some explanation beyond what supports its empirical adequacy. IBE goes further than the constructive empiricist's belief in empirical adequacy, and, according to van Fraassen, it is the latter that the realist also has, which is what actually enables science to continue. Van Fraassen rejects realism not because he thinks it irrational but because he rejects the 'inflationary metaphysics' that is an account of laws, causes, kinds and so on, which he thinks must accompany it. Empiricists should repudiate beliefs that go beyond what we can (possibly) confront with experience, and this restraint allows them to say 'good bye to metaphysics' (van Fraassen 1989: 480). He thinks constructive empiricism offers an alternative view that offers a better account of scientific practice without such extravagance (van Fraassen 1980: 73).

Van Fraassen suggests that to be an empiricist is to believe that 'experience is the sole source of information about the world' (van Fraassen 1985: 253), yet this doctrine does not itself seem to be justifiable by experience. However, he has argued in a recent paper that empiricism cannot be reduced to the acceptance of such a slogan and that empiricism is in fact a 'stance' in the sense of an orientation or attitude towards the world (van Fraassen 1994). I am not sure why adopting an unjustifiable stance or attitude is any more respectable than simply accepting a proposition like the one above, which cannot be justified on its own terms. In any case, constructive empiricism has no normative force for a non-empiricist and so it may seem as if we have reached stalemate.

7.3   Common sense, realism and constructive empiricism

In the face of van Fraassen's critique of IBE the realist may fall back to the following view expressed by John Worrall:

Nothing in science is going to *compel* the adoption of a realist attitude towards theories [. . .] But this leaves open the possibility that some form of scientific realism, while strictly speaking unnecessary, is nonetheless the *most reasonable* position to adopt.

(Worrall 1984: 67)

If there is no rule of inference like IBE, such that scientific realism is implied by minimal rationality, why is it the most reasonable position to adopt? Worrall argues that this issue is directly analogous to the defence of commonsense metaphysical realism against the sort of radical scepticism we discussed in Chapter 5. In order to make sense of our perceptions, we are not *compelled* to assume the existence of a real, external world; nonetheless, this seems the most reasonable position to take. So, the realist claims, the constructive empiricist must be epistemically erratic, because scientific realism is nothing more than the analogue of commonsense metaphysical realism in the unobservable domain.

Van Fraassen denies that there must be some explanation in terms of unobservables for the 'persistent similarities' in the phenomena (van Fraassen 1985: 298). However, realists argue that the only reason for accepting the objective existence of the table in front of me is to explain the persistent similarities in the phenomena. Therefore, it might seem that all the good arguments for the existence of tables carry over to the existence of electrons and, similarly, van Fraassen's arguments for withholding belief in the latter likewise motivate scepticism about the former. The belief in everyday objects allows us to explain many observable phenomena that would otherwise be inexplicable. Why should such explanation be ruled out for the unobservable world? Hence, many realists argue that van Fraassen ought really to be a sceptic about other minds and external objects, since these are all posited to explain the regularities in our experience.

Van Fraassen does not argue against commonsense realism and says that he is not a sceptic about 'tables and trees' (van Fraassen 1980: 72). However, whether or not he intends to argue against it, if his position in epistemology makes commonsense realism lack any justification then this is surely significant. It may be that constructive empiricism can meet this challenge, but van Fraassen owes us an argument. He seems to assume that commonsense realism is free of metaphysics, and that it is established by denying that there are sense data (van Fraassen 1980: 72). He is quite sure that the latter do not exist (after all, they are theoretical entities so it would hardly be consistent of him to believe in them). However, denying the existence of sense-data is not sufficient to establish metaphysical realism about the world of common sense that philosophers such as Devitt, Worrall and most other scientific realists take for granted. If constructive empiricism is founded on such a weak epistemology that it cannot safeguard the rationality of belief in, or at least the irrationality of denying, the existence of the commonsense world, then it is hardly surprising that it cannot do the same for the unobservable world of theoretical science. However, it has become apparent that van Fraassen's philosophical outlook is very different from that of his realist opponents. He wants to begin epistemological inquiry *in* the life-world, rather than taking its fundamental task to be to safeguard belief in the existence of the lifeworld. His perspective is like that of the phenomenologists who take the existence of themselves in a public world to be their starting point. From this perspective, everyday objects are manifest, hence one does not need to make inferences to believe in them.

It seems the debate between scientific realists and van Fraassen leads back to the debate about the existence of everyday objects that sometimes gets philosophy a bad press. In the next chapter, I will turn to arguments for antirealism motivated by empirical facts about science and its history.

————o☺o————

Alice: If we were all as cautious in our beliefs as you nobody would ever believe in anything other than what is right in front of them. You're just being a sceptic where it suits you and not otherwise. If you don't believe in atoms I don't see why you believe in anything at all.

Thomas: We've been over this. Atoms aren't observable – they are just hypothetical entities that scientists have invented to explain things.

Alice: You might as well say the same thing about the dark side of the Moon.

Thomas: Well, anyway, there is an extra problem with the things scientists tell us exist. What about all the past theories that we don't use any more. If you had been around a few hundred years ago you would have believed that atoms have little springs and wheels on them because that's what scientists used to say then.

Alice: Now you are just being silly.

Thomas: No, I'm serious. All sorts of things were supposed to exist according to past theories. Newton thought that light was made of particles and Einstein didn't believe in quantum mechanics so how do you know that the scientists now won't be proved wrong in the future?

Alice: I don't believe that theories in modern science have changed that much, and even if they have nobody is saying that we ought to believe in every bit of every theory that we have now, just that most of our theories are more or less right.

Thomas: How can you be more or less right about the big bang? It either happened or it didn't, and light is either particles or waves – it can't be both.

Alice: Actually I think it is.

Thomas: That doesn't make any sense. You can't have it both ways. Either current science is right or wrong about atoms, the big bang and so on. I think that on past evidence we are safe to assume that in the future most of it will be replaced by new theories.

Further reading

Explanation

Friedman, M. (1974) 'Explanation and scientific understanding', *Journal of Philosophy*, **LXXI**, pp. 5–19.

Hempel, C. (1965) *Aspects of Scientific Explanation*, New York: Free Press.

Nagel, E. (1961) *The Structure of Science: Problems in the Logic of Scientific Explanation*, New York: Harcourt, Brace and World.

Ruben, D. (1990) *Explaining Explanation*, London: Routledge.

Ruben, D. (ed.) (1993) *Explanation*, Oxford: Oxford University Press.

Salmon, W. (1984) *Scientific Explanation and the Causal Structure of the World*, Princeton, NJ: Princeton University Press.

Van Fraassen, B.C. (1980) *The Scientific Image*, Oxford: Oxford University Press.

Laws of nature

Armstrong, D. (1983) *What is a Law of Nature?*, Cambridge: Cambridge University Press.

Van Fraassen, B.C. (1989) *Laws and Symmetry*, Oxford: Oxford University Press.

Inference to the best explanation

Ladyman, J., Douven, I., Horsten, L. and van Fraassen, B.C. (1997) 'In defence of van Fraassen's critique of abductive reasoning: a reply to Psillos', *Philosophical Quarterly*, 47, pp. 305–321.

Lipton, P. (1991) *Inference to the Best Explanation*, London: Routledge.

Psillos, S. (1999) *Scientific Realism: How Science Tracks Truth*, London: Routledge.

Van Fraassen, B.C. (1989) *Laws and Symmetry*, Oxford: Oxford University Press.

8

---•◦☉◦•---

Realism about what?

It is tempting to dismiss worries about the underdetermination of theories as purely philosophical doubts. Of course we cannot prove beyond all doubt that electrons exist, but then we cannot prove beyond all doubt that the Sun will rise tomorrow, that all metals expand when heated, or that everyday objects such as tables are still there when we aren't perceiving them. Hence, assuming that the mere possibility of error is not sufficient seriously to undermine ordinary claims to know, neither does it threaten scientific realism. Indeed, as we saw in the previous chapter, the very most that van Fraassen can claim to have shown is that it is not irrational for us to withhold belief in unobservables, but neither is it irrational to believe in them, and this is much weaker than traditional forms of scepticism about science. In this chapter, we will consider arguments for various kinds of antirealism, which are motivated by careful scrutiny of the practice and history of science, rather than by epistemological scruples. In different ways, facts about real science raise the question 'what should we be realists about?'.

8.1   Theory change

Perhaps the most compelling argument against scientific realism is the notorious 'pessimistic meta-induction', which in recent years has been championed by Larry Laudan (1981). This is a form of 'argument from theory change', and differs from the underdetermination argument by appealing to history, rather than to theoretical possibilities and baroque mathematical constructions. Recall the (global IBE) argument of 7.2.2; roughly, there is an explanatory connection between the empirical success of scientific theories and their truth, in virtue of which scientific realism offers the only, or at least the best, explanation of the progress of science. Laudan turns this argument around and argues that we have positive reason, by induction, for not believing in the existence of the theoretical entities described by our best current theories. Like the global defence of realism, this argument is at the 'meta-level' because it is premised on consideration of science as a whole and its evolution in time. Laudan produced a list of now abandoned theories that once enjoyed predictive and explanatory success. Many of these theories featured theoretical terms, such as 'phlogiston', 'ether' and 'caloric', which were supposed to refer to various kinds of unobservables but which, according to modern science, fail to refer to anything at all (just as the terms 'unicorn', 'dragon' and 'leprechaun' don't refer to anything). If this is so then, rather than being justified in believing our best current theories are true, we have inductive grounds for the pessimistic conclusion that they too are likely to be replaced by successors which will show them to be false. Consequently, we have good reason to believe that terms like 'electron' don't refer to unobservable particles of matter after all. Hence, whatever the merits of IBE and whether or not constructive empiricism is ultimately defensible, scientific realism cannot be the best explanation of the success of science because it is not even empirically adequate. Notice that this is an argument for *atheism* about unobservable entities – believing that electrons and the like don't exist – not van Fraassen's type of *agnosticism*.

8.1.1 Approximate truth

For a while, in the eighteenth and nineteenth centuries, it seemed as if Newtonian mechanics might offer a completely true account of the behaviour of matter. It was widely thought that light was composed of particles and therefore that Newton's laws of motion must underlie optics too. Indeed, when Maxwell's electromagnetism became widely accepted many physicists thought that this in no way impugned the status of Newtonian theory because they assumed that somehow the former would eventually be reduced to the latter. The confidence in classical physics was such that, at the end of the nine-teenth century, some physicists apparently thought that their subject was nearly finished and only a few problems remained to be ironed out. Even the philosophically sophisticated Poincaré, whose know-ledge of contemporary science and its history was unrivalled, was convinced that Euclidean geometry and Newtonian mechanics would always reign supreme. However, subsequently, the discovery of radioactivity opened up a whole new domain of inquiry and eventu-ally led to the positing of two new fundamental forces (the strong and weak nuclear forces), where classical physics recognised only gravity and electromagnetism as fundamental. Furthermore, of course, Einstein's theories of special and general relativity forced a major revolution in the understanding of gravity, space and time. Similar revolutions have taken place in chemistry; for example, the adoption of the modern theory of atomic structure; and in biology, for example the acceptance of evolution and then DNA. Hence, nowadays, even those who have great faith in science and never question scientific realism are not so naïve as to think that all the best contemporary scientific theories are wholly true and without fault. There have been enough cases of successful theories being amended in unexpected ways, or of unforeseen new phenomena being discovered, for it to be apparent that even the finest science is subject to revision and correction.

Of course, this is partly what motivated Popper to think that all theories are merely conjectures rather than certitudes, but, as we saw in Chapter 3, he abandoned the idea that we ever have any positive reason to believe theories on the basis of evidence. On the other hand, most scientific realists have a conception of evidential support which entails that the predictive and explanatory success of a theory can provide inductive grounds for believing it to be true. They try and accommodate the lessons of history by claiming that theories are not perfectly but only 'approximately true'. Approximate truth, which is sometimes called *verisimilitude*, is indispensable to contemporary scientific realists, but despite a good deal of work it seems to have eluded a satisfactory, precise characterisation. Popper attempted a formal definition but it famously failed (see Psillos 1999: 261–264). Subsequently, people have developed formal mathematical and logical theories of approximate truth which they then apply to scientific theories. However, none of these is without its difficulties, and verisimilitude continues to puzzle philosophers.

Even though the idea of approximate truth and the idea of different theories being more or less approximately true than each other are very difficult to explicate, they are arguably needed for an adequate semantics of other types of propositions, not just theoretical scientific ones. For example, if I say that it is midday when it is actually 2 minutes past noon, then my utterance is strictly speaking false, and yet it is undeniably approximately correct. Where we are talking about numbers, the notion of approximate truth is very clear, because we know what it means for one number to be relatively close to another one. In such cases we can even quantify the degree of approximation. However, even when no numbers are involved, a notion of approximate truth may still be intelligible. For example, suppose someone declares that it is raining outside, but in fact it is sleeting or snowing. Clearly, what was said was true in so far as there is precipitation occurring, although false in so far as the form of the precipitation is different from what was stated. Similarly, it is approximately true that the sea is blue although it is usually tinged with green, and so on.

In science, examples of approximation and verisimilitude abound. Every schoolchild learns that the Earth is not flat but spherical, and we picture it as a globe, but of course, the surface of the Earth is very uneven with mountains and valleys, and furthermore it is slightly flattened at the poles. Strictly speaking then, it is not a sphere at all, yet we continue to describe it as such because that is close to the truth in salient respects. Similarly, if modern science is to be believed, theories such as Newtonian mechanics and Maxwell's electrodynamics are false, but scientific realists argue that they were, and are, approximately true, although less so than contemporary theories. Of course, to some extent it is approximately true that the Earth is flat, because it is so big relative to us that for many purposes the degree of curvature is negligible. The danger with the idea of verisimilitude is that it may inherit the permissiveness and relativism of similarity in general. Any two objects are similar in some respect. Take, for example, the Sun and my kettle; they are similar to each other because they are both further from the centre of the Earth than the bottom of the Pacific ocean is. Judgements of similarity can only be evaluated relative to a pragmatic context. For example, a glass tumbler is more similar to a cup than to a lens relative to function, but more similar to a lens than to a cup relative to constitution. So, if all that realists claim is that the best contemporary scientific theories are similar to the truth, this might be a very weak claim. (Antirealists may argue that the notion of approximate truth is in danger of collapsing onto the notion of approximate empirical adequacy. To say that the predictions of theories for what we can observe are approximately accurate does not amount to a defence of realism.)

Nonetheless, since it is difficult to see how we could get by without some notion of verisimilitude in analysing discourse about the observable world, it would be unreasonable to reject scientific realism, in particular, merely on the grounds that the notion of approximate truth is somewhat vague. In any case, as we shall see, in an attempt to give some content to the claim that a theory is approximately true, many realists have focused on the idea of successful reference. A term can be said *to refer* successfully if there is something, or some things, which are picked out by it. So, for example, 'horse' refers but 'the present king of France' does not. In his challenge to realism, Laudan (1981) argues that a scientific theory cannot be even approximately true if its central theoretical terms do not refer to anything. Hence, in the debate about the extent to which theory change in the history of science motivates scepticism about unobservables, it is often the referential status of abandoned theoretical terms that is in dispute.

8.1.2   *Sense and reference*

Analytic philosophers often distinguish between the *sense* of a term and its *reference*. Sense is a matter of the ideas and descriptions associated with the term, whereas the reference is the thing or things the term is used to talk about. Sense and reference are, of course, related, but they are not the same. If they were it would be impossible for us to continue to talk about the same thing if we changed our minds about its properties. The sense of the term 'whale' used to include the concept of a fish, whereas this has now been replaced by the concept of a mammal. Yet 'whale' still refers to the animals that our ancestors enthusiastically hunted. In the case of observable entities, reference can be fixed by pointing to the thing or things in question. This is what philosophers call 'ostensive definition'. This will work even when the sense of the term is at odds with the properties of the thing to which it refers. For instance, someone might say 'Audrey is the one over there talking to Angela'. Suppose that the former woman is in fact talking to someone else, but both parties in the conversation falsely believe her interlocutor to be Angela. Then the description associated with the introduction of the term 'Audrey' will fail to pick out Audrey correctly, yet both people can still successfully refer to her.

The problem is that theoretical terms in science that are alleged by realists to refer to unobservable entities cannot be given an ostensive definition. Instead, it is natural to think that the reference of a term like 'electron' is fixed by the theory of electrons. Therefore, it refers to very small entities that are negatively charged, orbit the nuclei of atoms, have a particular mass, and so on. This amounts to saying that the sense of such terms fixes their reference, and yet, as we saw in Chapter 4, Kuhn pointed out that the sense of many scientific terms such as 'atom', 'electron', 'species' and 'mass' has changed considerably during the course of scientific revolutions. If the reference of theoretical terms is fixed by the whole of the theories in which they feature, then any change in the latter will result in a change in the former. It seems as if we must either accept that the world itself changes when theories change, or that most theoretical terms don't refer to anything.

In response to Kuhn, Putnam (1975b) proposed a radically different account of the meaning of theoretical terms. He pointed out that most of us have no idea how many terms get their reference, but we nonetheless successfully refer to particular kinds of things with them. For example, which things in the world are referred to by the terms 'gold', 'elm' or 'French spaniel' is fixed by a few experts, to whom everyone else defers, yet such terms are part of a common language which we all use despite our ignorance. It is not just that most of us lack an explicit definition, we also often lack any way of distinguishing samples. The experts, on the other hand, have detailed criteria, often requiring subtle empirical tests; for example, in the case of gold these would include weight, reaction with particular acids, electrical resistance, melting point and so on. Putnam calls this *the division of linguistic labour*.

According to Putnam's theory of meaning for kind terms there are four components. Take the term 'water' for example. First there is its *syntactic marker*, which is that of 'mass noun' (like 'air' or 'wood', as opposed to count nouns, like 'horse' or 'tree'). Then there is the *semantic marker*, which is the idea of a common liquid, and the *stereotype*, which is that water falls from the sky as rain, is drinkable, transparent and so on. Finally, there is the *extension*, which is the actual stuff, the natural kind H_2O, to which 'water' refers. Putnam advocates a 'causal theory of reference' for natural kind terms such as 'water', 'gold' and 'electron', according to which 'water' is whatever causes the experiences that give rise to water talk. Reference is fixed not by the description associated with a term but by the cause that lies behind the term's use. This theory allows for continuity of reference across theory changes. Scientific theories help fix the stereotypes of terms such as 'electron', even though theories about electrons have changed, and hence the meaning of the term has changed, Putnam argues that 'electron' has always referred to whatever causes the phenomena that prompted its introduction, such as, for example, the conduction of electricity by metals. (In the nineteenth century, 'electron' was introduced to refer to the least unit of electric charge; however, physicists now believe that some quarks have a charge that is one third of the magnitude of the charge of electrons.) According to Putnam's theory, the reference of theoretical terms can be very stable indeed; the worry, as we shall see, is that it makes successful reference too easy to achieve.

8.1.3 The pessimistic meta-induction

The meta-induction has the following structure:

(i) There have been many empirically successful theories in the history of science that have subsequently been rejected and whose theoretical terms do not refer according to our best current theories.

(ii) Our best current theories are no different in kind from those discarded theories and so we have no reason to think they will not ultimately be replaced as well.

(iii) By induction, we have positive reason to expect that our best current theories will be replaced by new theories, according to which some of the central theoretical terms of our best current theories do not refer.

Therefore, we should not believe in the approximate truth or the successful reference of the theoretical terms of our best current theories.

In support of premise (i) Laudan lists the following theories, all of which he claims were empirically successful, but which, if our best contemporary theories are to be believed, have theoretical terms that do not refer:

the crystalline spheres of ancient and medieval astronomy;
the humoral theory of medicine;
the effluvial theory of static electricity;
the 'catastrophist' geology, with its commitment to a universal (Noachian) deluge;
the phlogiston theory of chemistry;
the caloric theory of heat;
the vibratory theory of heat;
the vital force theories of physiology;
the electromagnetic ether;
the optical ether;
the theory of circular inertia; and
theories of spontaneous generation.

(Laudan 1981: 29)

Laudan contends that his list could be extended 'ad nauseam' (Laudan 1981: 29)

However, the first realist strategy for dealing with the meta-induction is to restrict the number of theories that can legitimately be used as the basis for an induction whose conclusion concerns contemporary theories. It would be absurd to claim that we should be realists about any scientific theory, only those with the right features will qualify. One of the most common reactions to the list above is to complain that many of the theories mentioned are nothing like the best contemporary theories. Hence, although premise (i) above seems undeniable, premise (ii) is not.

(1)   Realist responses

(a) Restrict realism to *mature* theories

Most of Laudan's examples have not been taken seriously by realists. They reduce the number that make up the inductive basis, by only considering 'mature' theories from the past as relevant to contemporary science. A science reaches maturity when it satisfies requirements such as coherence with the basic principles of theories in other domains, and possession of a well-entrenched set of basic principles which define the domain of the science and the appropriate methods for it, and limit the sorts of theories that can be proposed. All parts of contemporary physical science incorporate the law of conservation of energy, as well as the basic theory of the structure of matter in terms of elements such as hydrogen, oxygen, carbon and so on. They employ a common system of units such as metre, kilogram, amp, volt, degree, and so on, and they all use concepts such as force, velocity, mass, charge and so on. Furthermore, theories in one domain are routinely deployed as background theories in other domains. Hence, it is arguable that contemporary science has a degree of unification and coherence, as well as mathematical sophistication, that is quite absent in many of theories Laudan cites.

(b) Restrict realism to theories enjoying *novel* predictive success

Many realists have argued that Laudan's notion of empirical success is much too permissive, and so they defend realism against the meta-induction by placing constraints on the kind of empirical success that justifies realism about a theory. After all, the right empirical consequences may be merely written into almost any theory in an ad hoc manner, yet we would not want to count such a theory's empirical success as evidence for its truth, not least because we could then be justified in believing each of two theories that attribute different causal structures to the world to be true. For example, for a time, the wave and particle theories of light both had a good degree of predictive and explanatory success in relation to known types of phenomena, yet they are ontologically incompatible. Many realists seem to have concluded that a theory needs to have not just empirical success, but to have yielded confirmed *novel* predictions.

The idea of novel predictions was celebrated by Popper (see 3.1). He contrasted the risky predictions of physics with the vague predictions of psychoanalysis, but he also wanted to justify the failure of scientists to abandon Newtonian theory when it was known to be incompatible with certain observations. As we saw in 3.5(5), often various modifications to background assumptions are made to try and accommodate observed facts that would otherwise refute established theories. Popper, and following him Lakatos and others, argued that this course of action is acceptable only when the new theory produces testable consequences other than the results that motivated it. So, for example, the postulation of a new planet to accommodate the observed orbit of a familiar one is legitimate because it ought to be possible to observe the former (or at least its effects on other bodies).

There is a general issue here concerning the confirmation of theories by evidence. Is when a particular piece of evidence was gathered relevant to the degree to which it confirms a hypothesis? *Predictionists* say that only new evidence confirms theories, whereas *explanationists* say that only the explaining of known facts confirms theories. Other philosophers think that both old and new evidence can confirm theories but that new evidence is particularly compelling. Similarly, it is the ability of theories to predict previously unsuspected results that many realists think is particularly miraculous on the supposition that realism is false. To see why, consider the following famous example.

In the early nineteenth century, both the wave and particle theories of light offered competing accounts of familiar optical phenomena. A ray of light directed at a plane mirror at a given angle will be reflected at the same angle (the law of reflection), which is just how a ball behaves when shot at the wall of a billiard table. Hence, the established laws of mechanics, in particular the conservation of momentum, used to explain the behaviour of material objects such as billiard balls, could also be used to the explain the law of reflection in terms of the particle theory of light. On the other hand, other phenomena were more naturally accommodated by the wave theory. The famous two-slit experiment showed that, in certain circumstances, light passing through two slits and hitting a screen would create a pattern of equally spaced light and dark bands, rather than two illuminated patches opposite the slits and shadow in between. This was explained by wave theorists in terms of light waves overlapping, and so adding to or cancelling each other's intensity at different points in space.

In 1818, Fresnel developed a mathematical theory according to which light consists of waves of a particular kind, namely *transverse* waves. A transverse wave is one that oscillates perpendicularly to its direction of motion, for example a wave on the surface of water. Fresnel's theory made derivations of known effects precise and elegant, but it had a new consequence. It predicted that, in certain circumstances, light that was shone on a completely opaque disk would cast a shadow with a bright white spot in its centre. The phenomenon is known as *conical refraction* and has now been observed many times. However, Fresnel knew nothing of it when he developed his theory, and indeed did not even derive the result himself. This is more striking than the prediction of the existence of an extra planet, because it is a prediction of a completely new and unexpected type of phenomenon. (Another good example – mentioned briefly in 3.1 and again in 6.1.3 – is the prediction of the general theory of relativity that light will take a curved path in the presence of very massive objects such as stars.)

(2) *Novelty*

If the meta-induction is to be defeated by restricting the theories that can form the inductive base to mature theories that enjoyed novel predictive success, the notion of novelty employed needs to be explained precisely. The most straightforward idea is that of *temporal novelty*. A prediction is temporally novel when it is of something that has not yet been observed. The worry about using this kind of novel predictive success as a criterion for adopting realism about particular theories is that it seems to introduce an element of arbitrariness into which theories are to be believed. When exactly in time someone first observes some phenomenon entailed by a theory may have nothing to do with how and why the theory was developed. It is surely not relevant to whether a prediction of a theory is novel or not, if it has in fact been confirmed by someone independently who didn't tell anyone about it. (As it turns out, the white spot phenomenon had been observed independently prior to its prediction by Fresnel's theory.) A temporal account of novelty would make the question of whether a result was novel for a theory a matter of mere historical accident and this would undermine the epistemic import that novel success is suppose to have for a particular theory.

It is more plausible to argue that what matters in determining whether a result is novel is whether a particular scientist knew about the result before constructing the theory that predicts it. Call this *epistemic novelty*. The problem with this account of novelty is that, in some cases, the fact that a scientist knew about a result does not seem to undermine the novel status of the result relative to the theory, because the scientist may not have appealed to the former in constructing the latter. For example, many physicists regarded the success of general relativity in accounting for the well-known, previously anomalous orbit of Mercury as highly confirming, because the reasoning that led to the theory appealed to general principles and constraints that had nothing to do with the empirical data about the orbits of planets. The derivation of the correct orbit was not achieved by putting in the right answer by hand.

Worrall suggests realism is only appropriate in the case of: 'theories, designed with one set of data in mind, that have turned out to predict entirely unexpectedly some further general phenomenon' (Worrall 1994: 4). However, in a recent analysis, Leplin rejects this as relativising novelty to the theorist and thereby introducing a psychological and so non-epistemic dimension to novelty (Leplin 1997). Take the case of Fresnel. If we say that the fact that the white spot phenomenon was known about is irrelevant, because Fresnel was not constructing his theory to account for it but it still predicted it, then we seem to be saying that the intentions of a theorist in constructing a theory determine in part whether the success of the theory is to be counted as evidence for its truth. Leplin argues that this undermines the objective character of theory confirmation needed for realism.

This motivates the idea of *use novelty*. A result is use-novel if the scientist did not explicitly build the result into the theory or use it to set the value of some parameter crucial to its derivation. Hence, Leplin proposes two conditions for an observational result O to be novel for a theory T.

> *Independence condition*: There is a minimally adequate reconstruction of the reasoning leading to T that does not cite any qualitative generalization of O.
>
> *Uniqueness condition*: There is some qualitative generalization of O that T explains and predicts, and of which, at the time that T first does so, no alternative theory provides a viable reason to expect instances.
>
> (Leplin 1997: 77)

Leplin explains that a reconstruction of the reasoning that led to a theory is an idealisation of the thought of the theorist responsible for that theory, and is said to be 'adequate' if it motivates proposing it, and 'minimally' so if it is the smallest such chain of reasoning (Leplin 1997: 68–71).

According to the two conditions above, novelty is a complex relation between a theory, a prediction or explanation, the reconstruction of the reasoning that led to it, and all the other theories around at the time, since the latter are required not to offer explanations of the result. It follows that: (a) if we found a dead scientist's revolutionary new theory of physics, but they left no record of what experiments they knew about or what reasoning they employed, such a theory could have no novel success. Hence, no amount of successful prediction of previously unsuspected phenomena would motivate a realist construal of the theory; (b) suppose that we already knew all the phenomena in some domain. In such a case we could never have evidence for the truth rather than the empirical adequacy of any theories that we constructed in this domain, no matter how explanatory, simple, unified or whatever. These consequences are surely contrary to realist thinking. Certainly scientific methodology includes far broader criteria for empirical success, such as providing explanations of previously mysterious phenomena. Indeed, Darwin's theory of evolution and Lyell's theory of uniformitarianism (according to which geological change is the result of ordinary forces acting over very long periods of time rather than sudden and dramatic catastrophes) were accepted by the scientific community because of their systematising and explanatory power, and in spite of their lack of novel predictive success. Realists also often argue that the unifying power of theories, which may bring about novel predictive success but need not do so, be taken as a reason for being realistic about them.

Furthermore, Leplin admits that his analysis makes novelty temporally indexed (Leplin 1997: 97), but this seems to fall foul of his own claim that 'it is surely unintuitive to make one's epistemology depend on what could, after all, be simple happenstance' (Leplin 1997: 43). The main problem seems to be with the uniqueness condition since it seems to leave too much to chance. For example, suppose a result is novel with respect to some theory, but that another theory comes along soon afterwards which also explains it. According to Leplin's view, a realist commitment to the former theory is warranted but not to the latter, yet it seems that the order of things might have been reversed, so that which theory is believed to be true is to some degree historically contingent. Moreover, truth is being imputed to explain the mystery of the novel success of one theory, but the success of the other theory, which would be novel were its rival not around, is left unexplained. This may leave us wondering whether the explanation of success in terms of truth is really necessary theory for either theory.

This does not mean that novel predictive success is a complete red herring. On the contrary, it does seem that the fact that theories sometimes produce predictions of qualitatively new types of phenomena, which are then subsequently observed, is a strong motivation for some sort of realism about scientific knowledge. However, it is the fact that novel predictive success is possible and happens at all that needs accounting for in terms of a general account of science and the world. On the other hand (like Psillos 1999), Leplin uses whether or not a theory enjoyed novel predictive success as a criterion for whether or not it, in particular, should be believed to be true, which leaves the fate of theories too much to chance.

8.1.4 Counter-examples to the no-miracles argument

Even if many of Laudan's examples are spurious, there are still some examples of theories that were both mature and that enjoyed novel predictive success. Even if there are only one or two such cases, the realist's claim that approximate truth explains empirical success will no longer serve to establish realism. This is because, where a theory is empirically successful, but is not approximately true, we will need some other explanation. If this will do for some theories then it ought to do for all, and then we do not need the realist's preferred explanation that such theories are true. Hence, we do not need to form an inductive argument based on Laudan's list to undermine the no-miracles argument for realism. Laudan's paper was also intended to show that the successful reference of its theoretical terms is *not* a necessary condition for the novel predictive success of a theory (Laudan 1981: 45), in other words, that there are counter-examples to the no-miracles argument:

(i) There are examples of theories that were mature and had novel predictive success, but whose central theoretical terms do not refer according to our best current theories.

(ii) Successful reference of its central theoretical terms is a necessary condition for approximate truth. (Premise (iv) of the meta-induction.)

(iii) There are examples of theories that were mature and had novel predictive success but which are not approximately true. (From (i) and (ii)).

(iv) Approximate truth and successful reference of central theoretical terms is not a necessary condition for the novel-predictive success of scientific theories.

(v) The no-miracles argument is undermined since, if approximate truth and successful reference are not available to be part of the explanation of some theories' novel predictive success, there is no reason to think that the novel predictive success of other theories has to be explained by realism.

Most attention has focused on the examples of the '*ether* theory of light' and the '*caloric* theory of heat'. Whatever the realists' preferred way of characterising novel predictive success, it seems these theories had it, so they need to account for these cases.

(1)   Realist responses to the counter-examples

There are two basic responses to the counter-examples.

(I) Develop an account of reference according to which the relevant abandoned theoretical terms refer after all

In 8.1.2 above we saw that realists have used causal theories of reference to account for the continuity of reference for terms such as 'atom' or 'electron', when the theories about atoms and electrons undergo significant changes. The difference with the terms 'ether' and 'caloric' is that they are no longer used in modern science. In the nineteenth century, the ether was usually envisaged as some sort of material, solid or liquid, which permeated all of space. It was thought that light waves had to be waves in some sort of medium and the ether was posited to fulfil this role. Yet if there really is such a medium then we ought to be able to detect the effect of the Earth's motion through it, because light waves emitted perpendicular to the motion of a light source through the ether ought to travel a longer path than light waves emitted in the same direction as the motion of the source through the ether. Various experiments failed to find such an effect. Furthermore, soon after Fresenel's ether theory of light had its great successes, Maxwell developed his theory of the electromagnetic field. Light was now regarded as a wave in the electromagnetic field, which was not considered to be a material substance at all. As a result, the term 'ether' was eventually abandoned completely.

However, the causal theory of reference may be used to defend the claim that the term 'ether' referred after all, but to the electromagnetic field and not to a material medium. If the reference of theoretical terms is to whatever causes the phenomena responsible for the terms' introduction, then since optical phenomena are now believed to be caused by the oscillations in the electromagnetic field, than the latter is what is referred to by the term 'ether'. Similarly, since heat is now believed to be caused by molecular motions, then the term 'caloric' can be thought to have referred all along to these rather than to a material substance. The worry about this is that it may imply that reference of theoretical terms is a trivial matter, since as long as some phenomena prompt the introduction of a term it will automatically successfully refer to whatever is the relevant cause (or causes). Furthermore, this theory radically disconnects what a theorist is talking about from what they think they are talking about. For example, Aristotle and Newton could be said to be referring to geodesic motion in a curved spacetime when, respectively, they talked about the natural motion of material objects, and the fall of a body under the effect of the gravitational force. We shall return to this issue below.

(II) Restrict realism to those theoretical claims about unobservables that feature in an *essential* way in the derivation of novel predictions

The essence of this strategy is to argue that the parts of theories that have been abandoned were not really involved in the production of novel predictive success. Philip Kitcher says that: 'No sensible realist should ever want to assert that the idle parts of an individual practice, past or present, are justified by the success of the whole' (Kitcher 1993: 142).

Similarly, Psillos argues that history does not undermine a cautious scientific realism that differentiates between the evidential support that accrues to different parts of theories, and only advocates belief in those parts that are essentially involved in the production of novel predictive success. This cautious, rather than an all or nothing, realism would not have recommended belief in the parts of the theories to which Laudan draws attention because, if we separate the components of a theory that *generated* its success from those that did not, we find that the theoretical commitments that were subsequently abandoned are the *idle* ones. On the other hand, argues Psillos: 'the theoretical laws and mechanisms that generated the successes of past theories have been retained in our current scientific image' (Psillos 1999: 108).

Such an argument needs to be accompanied by specific analyses of particular theories that both (a) identify the essential contributors to the success of the theory in question, and (b) show that these were retained in subsequent developments.

Psillos's strategy for defeating the argument from theory change is a combination of (I) and (II). Laudan claims that if current successful theories are approximately true, then the caloric and ether theories cannot be because their central theoretical terms do not refer (by premise (ii) above). Strategy (I) accepts premise (ii) but Psillos allows that sometimes an overall approximately true theory may fail to refer. He then undercuts Laudan's argument by arguing that:

- Abandoned theoretical terms that do not refer, such as 'caloric', were involved in parts of theories not supported by the evidence at the time, because the empirical success of caloric theories was independent of any hypotheses about the nature of caloric.
- Abandoned terms that were used in parts of theories supported by the evidence at the time do refer after all; 'ether' refers to the electromagnetic field.

Below, I identify some problems with this type of defence of realism and the claims made about 'ether' and 'caloric'. Most importantly, I argue that the notion of 'essential', deployed in (a) above is too vague to support a principled distinction between our epistemic attitudes to different parts of theories. In my view, ultimately the problem with strategy (II) is that it is in danger of becoming ad hoc and dependent on hindsight.

According to Psillos: 'Theoretical constituents which make essential contributions to successes are those that have an indispensable role in their generation. They are those which "really fuel the derivation"' (Psillos 1999: 110). This means that the hypothesis in question cannot be replaced by an independently motivated, non ad hoc, potentially explanatory alternative. (Remember that the sort of success referred to here is novel predictive success.) Psillos gives as an example of an idle component of a theory Newton's hypothesis that the centre of mass of the universe is at absolute rest. However, within Newton's system this hypothesis cannot be replaced by an alternative that satisfies the above requirements. That the centre of mass of the universe should move with any particular velocity is surely more ad hoc than it being at rest. There is also a sense in which the universe being at rest is simpler than it having a particular velocity, since the latter would raise the further question of what force set it in motion, requiring a new theory to explain its motion. Certainly, any alternative hypotheses that might have been entertained would not be explanatory of anything nor independently motivated. It seems that this hypothesis does count as an essential contribution to the success of Newton's theory, by Psillos' criteria, and hence that he would have had us be realistic about it. However, the notions of absolute space and absolute rest have no meaning in modern physics, so Psillos' criterion above has accidentally bolstered historically inspired scepticism.

This case is arguably not so serious for the realist for it does not involve a central theoretical term that was essentially deployed, yet which cannot be regarded as referring from the standpoint of later theory. Nonetheless, Psillos is intending to deal with the threat of such examples using this distinction between essential and inessential theoretical constituents, and this example reveals that his definition of this distinction does not, in general, capture only the theoretical hypotheses with which the realist would be happy. Another problem is the ambiguity concerning the type of dependence in question when we ask if a theory's success is dependent on a particular hypothesis. We can understand this as at least either logical/mathematical dependence or causal dependence. So, when we are asked to look at the particular novel empirical success of a theory and decide which are the parts of theory that this success depended on, then we will give different answers according to how we understand dependence.

Furthermore, the realist should be careful here for it is dangerous for realism overall to disconnect the metaphysical hypotheses and background assumptions about supposed entities, such as caloric or the ether, from what are construed as the real successes of the theories in issuing certain predictions. One of the central claims of contemporary realism is that we have to take seriously the involvement of theoretical and metaphysical beliefs in scientific methodology; that is, that we cannot disconnect the success of science from the theoretically informed methods of scientists. This is meant to support realism, for, according to Boyd and others, only realism explains why these extra empirical beliefs are so important. However, Psillos suggests that, after all, we need not take seriously scientists' beliefs about the constitution of the ether or caloric, because the success of the theories floats free of such assumptions. Let us now turn to the examples he discusses.

The case of the ether

Ether theories were successful by any plausible criteria that the realist may concoct and were mature, yet it was widely believed that the ether was a material substance, and there certainly is no such thing permeating space according to Maxwell's successor theory. The hypotheses about the material nature of the ether were no presuppositional or idle posits because they motivated the use of mechanical principles, such as Hooke's law, in investigations of how light waves would propagate in such a medium. This led to the fundamental departure from previous wave theories brought about by Fresnel's assumption that light propagates as a transverse wave motion. Hence, since Fresnel got important 'heuristic mileage' out of these mechanical principles, then, as Worrall (1995) has argued, although the replacing electromagnetic theory provides a constructive proof of the eliminability of the old theory, mechanics is really only eliminable from the success of Fresnel's theory in a minimal logical sense.

When we inquire into what hypotheses 'really fuel the derivation' we have no other way to address this question than by explaining how we would derive the prediction in question using our understanding of the theory. This does not show that hypotheses which *we* do not use in reconstructing the derivation played no role in making it possible for the scientists of the time to derive the prediction that they did. A modern scientist may not need to invoke anything about the material constitution of the ether to reconstruct Fresnel's predictions, but Fresnel did so to derive the predictions in the first place. Psillos says that 'essential constituents' of success are ones such that 'no other available hypothesis' (Psillos 1999: 309) can replace them in the derivation of the novel predictions made by the theory in question. The question is: *available to whom?* There was no other hypothesis available to Fresnel about the nature of the ether that would have allowed him to derive the novel predictions of his theory. In general, it seems true that quite often there will be no other hypothesis available at the time, but that in reconstructing derivations we may have several alternatives.

Psillos argues that there is continuity between the causal roles of attributes of the ether and those of the field. For example, the fundamental causal role of the ether is arguably to act as the repository of the energy associated with light between emission by a source and its absorption or reflection by matter. Light was known to travel at a finite velocity, so it had to be in some medium while passing through otherwise empty space. The electromagnetic field is now thought to be that medium. However, the selection of which causal role is important is done with the benefit of hindsight. Our assessment of what matters in the description of optical phenomena is very much relative to our current state of knowledge, as is any statement about the relevant causal role of some posited unobservable entity. However, we do not know what of current theories will be retained, in other words, what the *real* causal roles are.

It is true that the important principles about light (that it undergoes transverse propagation for example) are carried over into Maxwell's theory, and indeed there is a lot of continuity between the ether theories and electromagnetic field theory. However, the latter has now been replaced by quantum field theories, which may soon be replaced by a theory of superstrings or a grand unified theory of quantum gravity. It is implausible to suggest that 'ether' referred all along to a quantum field, because the latter has a completely different structure to Maxwell's electromagnetic field. For example, the latter is supposed to permeate all space and have a definite magnitude at different points, whereas the former is multi-dimensional and incorporates only probabilities for different magnitudes.

The case of caloric

In this case, Psillos argues that 'the approximate truth of a theory is distinct from the full reference of all its terms' (Psillos 1994: 161). Psillos argues that 'caloric' was not a *central* theoretical term, and we should only worry about the central theoretical terms: 'central in the sense that the advocates of the theory took the successes of the theory to warrant the claim that there are natural kinds denoted by these terms' (Psillos 1996: 312).

However, surely in the realism debate we ought only to be concerned with what scientists *should* believe and not with what they *do* in fact believe. If a group of antirealists came up with a novel predictively successful theory, realists would not admit that, because none of the scientists involved believes in the natural kinds denoted by the theory, the theory has no central terms. When we are concerned with philosophical disputation about science the fact that a particular scientist had this or that attitude to a theory is irrelevant. If it happened that all the heat theorists had been instrumentalists we would not then deduce that we should be instrumentalists.

In any case, Psillos argues that in the case of caloric theory, all the important predictive successes of the theory were independent of the assumption that heat is a material substance. He argues that all the well-confirmed content of the theory – the laws of experimental calorimetry – was retained in thermodynamics anyway. What I think Psillos is doing here is exploiting, as scientists often do, the positive side of the underdetermination problem. When we theorise about or model the phenomena in some domain we will inevitably make a few mistakes. If there was only one (the true) theory that could describe the phenomena, then we would almost certainly never hit upon it first time, and we usually need to modify theories in the face of new evidence. More importantly, after radical theory change we want to be able to recover the empirical success of old theories without buying into their outdated ontologies. The ever present possibility of alternative, empirically equivalent theories is therefore essential.

The problem is that the only reason that we do not say that 'caloric' refers (to what we call heat or thermal energy) is because it is regarded as essential to this term that it is a material substance. However, Psillos says that the scientists of the time did not commit themselves to this, that they withheld judgement about the nature of caloric. Since the reason for saying that 'ether' refers is that scientists were not committed to it being of any particular nature but were committed to its having some properties that, as a matter of fact, are carried over to Maxwell's theory, we could say then that 'caloric' (like 'ether') does refer. On the other hand, suppose that scientists had been so committed to belief in the material constitution of the ether that denying its material nature would seem as absurd as denying that caloric is a (fictional) material substance. Then, by the same argument as that for caloric, we would say that 'ether' does not refer to anything, since the derivations of the important predictions of ether theory are all independent of assumptions about its constitution (says Psillos), and that is the stated reason for saying that 'caloric' does not refer. Hence, whether we end up saying that this or that theoretical term refers to the same entities as a particular currently accepted theoretical term is arbitrary, and we could be in the position of saying that both 'ether' and 'caloric' did not refer.

It seems that even if Psillos' defence were to work, it would leave realism in a bit of a mess. Realists can no longer say that we should believe that the world is much like our best theories say it is, and that the theoretical terms of such theories genuinely refer. Instead, it is that some of these terms will refer, some not, and others will but only approximately. Furthermore, before realists commit themselves to any hypothesis or entity, they will have to look closely to see if it was essential in producing not just the success, but the novel predictive success of the theory. Where this is not the case they will not advise epistemic commitment. So, for example, before we know whether to believe that electrons are fundamental particles with spin 1/2 we had better check that their being spin 1/2 (and that their being fundamental and being particles) plays an essential role somewhere in the derivation of a novel prediction, and until this is done we had better be agnostic. On this account a theory can have great empirical success even though there is nothing in the world much like the entities postulated by the theory.

8.2 Multiple models

Roger Jones (1991) raises the problem 'Realism about what?'. He points to the existence of alternative formulations of physical theories that coexist in science, and argues that there is not necessarily a single picture of the world for us to believe in, even if we were assured of the epistemic warrant for doing so. For example, classical mechanics is a paradigm of a mature scientific theory and it has an impressive catalogue of predictive success from the return of Halley's comet to the recent successful space flights (classical mechanics is still used for such projects). The theory is first presented in terms of Newton's three laws of motion, understood as descriptive of a particle's behaviour. Velocity and acceleration are introduced as functional relationships between a particle's positions at different times that are directly measurable. The theoretical concepts of mass and force are defined operationally: the ratio of the masses of two particles is a constant of proportionality that, together with their initial velocities and accelerations, determines the relative motions of the two particles after they have interacted; force is defined as simply the product of mass and acceleration. (This gives us a problem, for a particle that never interacts with anything else could have any value whatever for its mass, but since real particles will always interact with something or other let us ignore this.)

Classical mechanics is applied to planetary phenomena by combining the law of gravitation with the three laws of motion. This approach has a great deal of success in dealing with large approximately spherically symmetrical bodies whose mass can be conveniently regarded as zero everywhere except at the centre of mass. However, for extended bodies, in general it proves computationally intractable, and is replaced by the field theoretic formulation of classical mechanics. In this approach, the fundamental object is the gravitational potential, which is defined for all points of space and the magnitude and direction of the potential gradient at a point is what determines (with the mass) the magnitude and direction of the force experienced by matter at that point.

So far then, we have two different ways of formulating classical mechanics; either as a theory of forces acting at a distance between point particles, or as a theory describing a gravitational field that occupies all points of space and acts locally. There is a third formulation to consider, in which the force law and the laws of motion can be derived from so-called minimum principles. This is often known as 'analytic mechanics' and was first developed by Euler and Lagrange and given a full treatment by Hamilton in the nineteenth century. Analytic mechanics is derived by invoking the 'principle of least action', which states that the path followed by a particle between two points will be such as to minimise the so-called 'action integral', which represents the total energy, kinetic and potential, of the system. The last way to formulate classical mechanics, to which Jones draws attention, is the curved spacetime formulation analogous to that of general relativity. The gravitational field is absorbed into the structure of spacetime and is represented by the degree of curvature at a given point.

Each of these approaches has similarities with other physical theories. The action at a distance approach was used for the formulation of the theory of the Coulomb force between charged particles in the late eighteenth century. The field-theoretic approach is directly analogous to the classical field theory of electromagnetism. Variational mechanics is related (mathematically at least) to general relativity and contemporary (gauge) field theories, and Hamiltonian mechanics is closely related to (non-relativistic) quantum mechanics. Suppose that our best theory of the world is classical mechanics and that matters of epistemology persuade us that we want to be realists; all this raises the question of what to be realistic about. The point is that each of these approaches suggests a quite different ontology and metaphysics. The first (Newtonian) approach suggests an ontology of point particles and forces acting at a distance. The field approach, on the other hand, suggests an ontology of action by contact in conformity with a principle of local causality, but postulates a new type of entity, the field. The curved spacetime approach suggests another picture again, where spacetime itself is part of the fundamental ontology of the theory and has a causal efficacy of its own. As for analytic mechanics, this seems to be independent of causal thinking, but requires instead a kind of teleology for particle behaviour according to which only properties of complete paths between points of space determine the motion of a body. The conclusion Jones draws from this is that the physicist, when asked to articulate the 'fundamental (theoretical) furniture of the Newtonian universe', will not be able to do so (Jones 1991: 190). According to Jones, classical mechanics is a 'vastly over-connected structure of concepts', which save the phenomena with very different explanatory frameworks and ontological commitments (Jones 1991: 190).

The short reply to Jones' problem would be just to point out that classical mechanics is known to be false, and that the existence of four alternative formulations of a false theory is no problem for the realist. However, this is an ultimately damaging response for the realist's position since it implies that they are only committed to a realist view of the one right theory of the world. Consider the situation at the end of the last century when Newtonian mechanics was widely held, on the basis of great predictive and explanatory success, to be true. If realism has any content it must allow that people were justified in being realists to some extent about classical mechanics; the question that Jones poses is 'which version?'. Of course, the curved spacetime formulation was not available until early in the twentieth century, but the problem remains in principle.

Alan Musgrave replies to Jones' argument and the points he makes would be endorsed by many realists. Jones presents the four theories above as 'versions of the same theory', but Musgrave denies this:

> Can these empirically equivalent Newtonian theories simply be different 'versions' or 'formulations' of one and the same theory? We will think so if we accept the positivist or antirealist or verificationist principle that empirically equivalent theories are really the same theory.
>
> (Musgrave 1992: 693)

Therefore the suggestion is that they are not, in fact, the same theory on any but an antirealist construal along positivistic lines; even van Fraassen concedes that theories that refer to different entities and posit different explanatory frameworks (as Jones says these do) should be taken literally, and therefore really are different.

However, the reason Jones offers for calling these four versions of classical mechanics different formulations of the same theory is that they are all taught to students as part of classical mechanics. As Musgrave concedes, Jones' examples are of strong empirical equivalence and are more interesting than usual cases of underdetermination because they are 'embedded in the actual practice of scientists in a way that the artificial examples concocted by philosophers are not' (Musgrave 1992: 693). In any case, Musgrave claims, like most realists, that empirical equivalence does not entail evidential equivalence because explanatory power is evidentially significant. Of course, each of the four theories Jones cites does have explanatory power, so none of them can be disregarded as easily as can the 'The world is as if T' (call this a surrealist transform of T). However, Musgrave's point is that once we deny that empirical equivalence implies evidential equivalence, we will not immediately assume that there are no rational grounds for choosing between the four theories:

Those who prefer causal to teleological explanation will rule out the minimum-principle 'version' of Newtonian mechanics. Does the history of science give us no reason to prefer causal to teleological explanation? Those who prefer local causality will rule out the action-at-a-distance 'version' of Newtonian mechanics. Scientists armed with a successful electromagnetic field theory might prefer the gravitational field theory because of the prospect it affords of unifying the two. And so forth.

(Musgrave 1992: 696)

Although antirealists see this intrusion of metaphysics into physics as unwarranted, realists such as Musgrave believe that physics and metaphysics are continuous; each informs the other:

Were Newton and many others foolish to worry about action-at-a-distance? Were Einstein and some others foolish to worry about the idealism that seemed implicit in the orthodox interpretation of quantum mechanics? Is it silly to oppose teaching our children creationist surrealism: 'God created the Universe in 4004 B.C. [or whenever] as if the teachings of natural science were true'?

(Musgrave 1992: 696)

Musgrave wants to persuade us that the only way to rule out (what we may agree is) the madness of creation science, Berkeley's idealism and so on, is by using metaphysical considerations that will also rule out Jones' multiplicity problem. However, Berkeleyian idealism is rejected because it requires an ontology of ideas that is discredited for a whole host of reasons, none of which will help us with classical mechanics. Furthermore, as already pointed out, nobody *uses* surrealist transforms of theories, which is another reason for rejecting them that we cannot invoke to answer Jones. Creation science is ad hoc and has contributed no new empirical success to biology or palaeontology, and so it does not have anything like the credibility of evolutionary biology. Musgrave characterises his theory of evidential support as 'partly historical' (Musgrave 1992: 695) and argues that the prediction of 'novel facts' is evidentially important. However, none of the disreputable theories he cites have had any novel predictive success whereas all the formulations of classical mechanics can claim some. It seems unwarranted to privilege the formulation of classical mechanics that happened to come first, when all of them could have predicted the same novel facts.

Even if we accept that explanatory power and metaphysical considerations are evidential, we have no easy solution to Jones' problem because different metaphysical criteria pull in different directions. Suppose that novel predictive success is our highest rated theoretical virtue; this would support the action-at-a-distance version of classical mechanics most. On the other hand, if we have a metaphysical penchant for local causality we will prefer the field theoretic version of the theory, which we will also prefer because of its compatibility with electromagnetic field theory. On the other hand, the variational approach to classical mechanics fits well with quantum mechanics, while the spacetime approach obviously coheres well with general relativity.

Musgrave ultimately claims: '[we] should be realists about what the best metaphysical considerations dictate, where the best metaphysical considerations are those that have yielded the best physics (Musgrave 1992: 691). However, there is no evidence that there is a unique set of metaphysical considerations that have yielded the best physics. On the contrary, the metaphysical presuppositions dominant at different stages in the history of science are themselves diverse and transient. Who would now argue that nature abhors a vacuum, or that all action must be contact between impenetrable particles, or indeed that a new theory cannot be adopted unless it admits of a mechanical reduction? Einstein was arguably influenced in the development of relativity theory by positivistic leanings, but surely this does not mean the realist should adopt positivism. There is also an apparent circularity in using metaphysical considerations to discover which is the best physics we have that then tells us what metaphysics to adopt.

8.3 Idealisation

The examples of scientific laws and theories used in philosophical debates about laws, explanation and so on are simplistic and perhaps misleading, given the chaos and multiplicity of the complex historical processes of theory development and transmission. Moreover, since philosophers often do not discuss theories unless they have already come to form part of received scientific knowledge, the theories they discuss are generally the product of rational reconstruction by scientists. This has led to a critique of the methods and results of philosophers of science by those who have noticed the lack of correspondence between science as the object of philosophical analysis and science as a part of the lifeworld. Nancy Cartwright allies herself with those who have downplayed the significance of theory as opposed to practice for the philosophy of science, and advocates a new kind of instrumentalism:

> Our scientific understanding and its corresponding image of the world is encoded as much in our instruments, our mathematical techniques, our methods of approximation, the shape of our laboratories, and the pattern of industrial developments as in our scientific theories. [...] these bits of understanding so encoded should not be viewed as claims about the nature and structure of reality which ought to have a proper propositional expression that is a candidate for truth or falsehood. Rather they should be viewed as adaptable tools in a common scientific tool box.
>
> (Cartwright *et al.* 1995: 138)

Let us accept that there is much more to science than theories. Nonetheless, given that theories play some role in encoding our scientific knowledge, we may still inquire into what exactly that role is. Cartwright makes her view plain: 'Physics does aim to represent the world, but it represents it not in its theories but in its models' (Cartwright *et al.* 1995: 139).

One of Cartwright's central claims is that the traditional view of theories and models does not reflect the actual scientific practice of applying theories. The traditional view in question is the 'covering law' account: it suggests that theories need only to be conjoined with the appropriate auxiliary assumptions for particular contexts, and that together with these they will imply the data/phenomena. The actual use of theories and the construction of models in science involves the application of abstract theoretical structures to concrete situations. Cartwright's distinctive metaphysics of science is motivated by her analysis of theory application. Her book, *How the Laws of Physics Lie*, made the distinction between *phenomenological* and *fundamental* laws and argued that the former refer to the concrete and the particular, and the latter to the abstract and the general. Furthermore: 'The phenomenological laws are indeed true of the objects in reality – or might be; but the fundamental laws are true only of the objects in the model' (Cartwright 1983: 4).

According to this book, the fundamental laws, because of their abstract nature, may be explanatory, but they do not describe what happens at all. They are linked to the appearances by phenomenological laws, which are non-explanatory but descriptive. Hence, at the theoretical level scientists construct models that are overtly of a sort that the real things do not fit. In order to relate these to specific phenomena they have to carry out a 'theory entry' process (Cartwright 1983: 132–134), whereby the phenomena are connected to theoretical models through a 'prepared description' that is overtly incorrect. This process is what we may loosely call 'idealisation'.

The most ubiquitous form of idealisation in physics is the application of mathematics to the physical world. For Duhem, this was enough to make the theoretical claims of physics quite separate from the ordinary truth-valued claims of everyday life, for the former are expressed in terms of concepts that are applied only with the help of artificially precise mathematics. Hence, he held that physical concepts are abstract and describe only imaginary constructions; they are mere symbolic formulae depending on a whole raft of theory for application to reality. Our concern here is not with 'mathematical idealisation', but with idealisations such as the point particle, the frictionless plane, the rigid rod and so on.

Cartwright distinguishes between cases where idealisation is performed on a concrete object/situation, and cases that are often called idealisation but where the simplifying assumptions involve abstracting, so that we are no longer dealing with concrete, but rather with abstract (and fictional), entities. For her, idealisation is the theoretical or experimental manipulation of concrete circumstances to minimise or eliminate certain features. For example, we idealise a real surface and get a frictionless plane, and then reintroduce a coefficient for friction with a convenient mathematical form. Here, the laws arrived at are approximately true, and in the laboratory it is possible to apply them directly to very smooth surfaces and so on. Thus, the laws arrived at by idealisation are still *empirical* or *phenomenological*, that is they are still about *concrete* situations.

In the case of abstraction, we subtract concrete facts about objects, including perhaps the details of their material composition, and – especially importantly – we eliminate interfering causes. Hence, says Cartwright, the laws arrived at cannot be even approximately true, since relevant causal features have been subtracted and the laws are therefore not about concrete situations. In the traditional view, these abstract or fundamental laws are genuine claims about reality. Cartwright has argued that these laws lie if we interpret them as telling us about regularities in concrete situations, and that they are in fact *ceteris paribus* laws, and thus not true of any actual or possible concrete situation. For example, the law of gravitation states what happens to bodies upon which no other forces are acting; but there are no such bodies in the actual universe, and so strictly speaking it cannot be true of anything.

8.4   Structural realism

[T]he structure of this physical world consistently moved farther and farther away from the world of sense and lost its former anthropomorphic character [. . .] Thus the physical world has become progressively more and more abstract; purely formal mathematical operations play a growing part.

(Planck 1996: 41)

As we have seen, in the debate about scientific realism, arguably the two most compelling arguments around are the 'no miracles' argument, and the 'pessimistic meta-induction'. In an attempt to break this impasse and have 'the best of both worlds', John Worrall introduced *structural realism* (although he attributes its original formulation to Poincaré) (Worrall 1989). Using the case of the transition in nineteenth century optics from Fresnel's elastic solid ether theory to Maxwell's theory of the electromagnetic field, Worrall argues that:

There was an important element of continuity in the shift from Fresnel to Maxwell – and this was much more than a simple question of carrying over the successful empirical content into the new theory. At the same time it was rather less than a carrying over of the full theoretical content or full theoretical mechanisms (even in 'approximate' form). [. . .] There was continuity or accumulation in the shift, but the continuity is one of form or structure, not of content.

(Worrall 1989: 117)

According to Worrall, we should not accept full blown scientific realism, which asserts that the *nature* of things is correctly described by the metaphysical and physical content of our best theories. Rather, we should adopt the structural realist emphasis on the mathematical or *structural* content of our theories. Since there is (says Worrall) retention of structure across theory change, structural realism both (a) avoids the force of the pessimistic meta-induction (by not committing us to belief in the theory's description of the furniture of the world), and (b) does not make the success of science (especially the novel predictions of mature physical theories) seem miraculous (by committing us to the claim that the theory's *structure*, over and above its empirical content, describes the world).

Although structural realism has attracted considerable attention, Worrall's proposal needs to be developed if it is to provide a substantive alternative to traditional scientific realism. In particular, there is a fundamental question about the nature of structural realism that should be answered: is it metaphysics or epistemology? Worrall's paper is ambiguous in this respect. At times his proposal seems to be that we put an epistemic constraint on realism to the effect that we only commit ourselves to believing in the structural content of a theory, while remaining agnostic about the rest. This is suggested by Poincaré who talks of the redundant theories of the past capturing the 'true relations' between the 'real objects which Nature will hide for ever from our eyes' (Poincaré 1905: 161).

On the other hand, Worrall's position is not explicitly an epistemic one, and other comments suggest a departure from the metaphysics of standard scientific realism. For example, he says: 'On the structural realist view what Newton really discovered are the relationships between phenomena expressed in the mathematical equations of his theory' (Worrall 1989: 122). If the continuity in scientific change is of 'form or structure', then perhaps we should abandon commitment to the reference of theories to objects and properties, and account for the success of science in other terms. Redhead says: 'realism about what? Is it the entities, the abstract structural relations, the fundamental laws or what? My own view is that the best candidate for what is "true" about a physical theory is the abstract structural aspect' (Redhead 1996: 2). This seems to concur with structural realist sentiments expressed by Howard Stein: 'our science comes closest to comprehending "the real", not in its account of "substances" and their kinds, but in its account of the "Forms" which phenomena "imitate" (for "Forms" read "theoretical structures", for "imitate", "are represented by")' (Stein 1989: 57).

Structural realism has been the subject of recent debate in the philosophy of physics too but I cannot discuss it further here. I hope the reader will by now be aware that the question of scientific realism is much more complex than it first appears. In order to take the discussion further we would have to consider in detail the history of science, the nature of our best contemporary theories, and the implications of idealisation and the practice of science.

Further reading

On the meta-induction

Hardin, C.L. and Rosenberg, A. (1982) 'In defence of convergent realism', in *Philosophy of Science*, **49**, pp. 604–15.

Kitcher, P. (1993) *The Advancement of Science: Science without Legend, Objectivity without Illusions*, Oxford: Oxford University Press.

Laudan, L. (1981) 'A confrontation of convergent realism', *Philosophy of Science*, **48**, pp. 19–48, reprinted in D. Papineau (ed.) (1996) *Philosophy of Science*, Oxford: Oxford University Press.

Psillos, S. (1999) *Scientific Realism: How Science Tracks Truth*, London: Routledge.

Worrall, J. (1989) 'Structural realism: the best of both worlds?', *Dialectica*, **43**, pp. 99–124, reprinted in D. Papineau (ed.) (1996) *Philosophy of Science*, Oxford: Oxford University Press.

On reference

Field, H. (1995) 'Theory change and the indeterminacy of reference', in P. Lipton (ed.) *Theory, Evidence and Explanation*, Aldershot: Dartmouth.

McCulloch, G. (1989) *The Game of the Name: Introducing Logic, Language and the Mind*, Chapters 2 and 3, Oxford: Oxford University Press.

Papineau, D. (1979) *Theory and Meaning*, Section 5.6, Oxford: Oxford University Press.

Putnam, H. (1975) 'The meaning of "meaning"', in his *Mind, Language and Reality: Philisophical Papers*, Volume 2, Cambridge: Cambridge University Press.

Theories and models

Cartwright, N. (1983) *How the Laws of Physics Lie*, Oxford: Oxford University Press.

Dupré, J. (1993) *The Disorder of Things: Metaphysical Foundations of the Disunity of Science*, Cambridge, MA: Harvard University Press.

Jones, R. (1991) 'Realism about what?', *Philosophy of Science*, 58, pp. 185–202.

Musgrave, A. (1992) 'Discussion: realism about what?', *Philosophy of Science*, 59, pp. 691–97.

Structural realism

Gower, B. (2000) 'Cassirer, Schlick and "structural" realism: the philosophy of the exact sciences in the background to early logical empiricism', *British Journal for the History of Science*, 8, pp. 71–106.

Ladyman, J. (1998) 'What is structural realism?', *Studies in History and Philosophy of Science*.

Glossary

—◦◯◦—

analytic   An analytic truth is a statement whose truth or falsity is determined solely by the meanings of the terms that make it up, such as 'all grandmothers are the mothers of one of their grand-children's parents', hence it expresses what is sometimes called a relation among our ideas. A synthetic statement is one that is not analytic such as 'all life on Earth is carbon-based'.

a priori   *A priori* knowledge is knowledge that is justified independently of any sensory experience. Traditionally, some philosophers have argued that mathematics and logic are subjects about which it is possible to have *a priori* knowledge. Empiricist philosophers deny that *a priori* knowledge of matters of fact is possible and try and account for knowledge of mathematics either by saying that such truths are all analytic (and so express mere relations among our ideas), or by arguing that mathematical knowledge is based on experience. *A posteriori* knowledge is that which is not *a priori* and hence is justified on the basis of sensory experience.

antirealism   In the philosophy of science, antirealism is any view that denies that we know that even our best scientific theories refer to mind-independent unobservable entities.

causal realism   This is the view that there are mind-independent or external objects, but that we only interact indirectly with them.

deduction   Deduction is inference in accordance with the laws of logic. A deductively valid argument or inference is one where it is not possible for the premises all to be true while the conclusion is false. A sound argument is one that is valid and where all the premises are true (and hence so is the conclusion).

demarcation problem The problem of providing a general rule or criterion for distinguishing science from non-science, and especially for distinguishing genuine science from activities or theories that are claimed to be scientific but which are not. The usual examples of pseudo-sciences given by philosophers and scientists are psychoanalysis and astrology.

direct realism The view that there are external objects that exist independently of our minds and which we directly perceive with the senses. Hence, this is a form of metaphysical realism that denies ideaism.

empiricism The term empiricism may be used for two related but distinct doctrines: the first is that all our concepts are derived in some way from sensory experience (concept empiricism), in other words there are no innate ideas; the second is that all knowledge of reality derives its justification from sensory experience, in other words there is no substantial *a priori* knowledge.

epistemic relativism Epistemic relativism is the view that what counts as knowledge, as opposed to merely true belief, is relative to the standards of some social group. Hence, knowledge is just those beliefs that are considered legitimate by particular institutions and authorities within some society.

epistemology The theory of knowledge or epistemology is that part of philosophy concerned with the nature, sources and justification of knowledge. Hence, philosophers speak of 'epistemological' problems, theories and so on. (This term is not to be confused with 'epistemic' which means pertaining to knowledge rather than to the theory of knowledge.)

falsificationism The theory of the scientific method originated by Popper and developed by Lakatos, according to which science is fundamentally about trying to falsify theories rather than trying to find evidence in their favour.

form A technical term in the philosophy of Plato and Aristotle meaning the structure or essence of a thing as opposed to its matter or substance. For example, the form of a statue is its shape, whereas its matter is the lump of marble from which it is made. The term 'form' is sometimes used in the history of science to refer to the real nature or cause of something.

foundationalism In epistemology the theory according to which our justified beliefs fall into two categories, namely basic beliefs, which are justified independently of all other beliefs, and non-basic beliefs, which are those that are justified by their inferential relations to basic beliefs. Foundationalism comes in different varieties depending on whether basic beliefs have to be certain or can be fallible.

hypothetico-deductivism   A theory of the scientific method according to which science proceeds by the generation of hypotheses, from which predictions are deduced that can be tested by experiment. This theory can be cashed out in inductivist or falsificationist terms depending on whether positive test results are regarded as confirming the theory or not, respectively.

ideaism   This is the view that the immediate or direct objects of sensory experience are not objects in the external world, but our own ideas (or representations or sense-data).

idealism   Idealism is the view that everything that exists is ultimately mental or spiritual in nature. Hence, idealists deny the existence of mind-independent material objects.

induction   In the broadest sense of the term, induction is any reasoning that is not deductively valid. In the narrow sense it refers to any inference from the past behaviour of things to their future behaviour, that is any inference from the observed to the unobserved.

inductivism   Any theory of the scientific method according to which generalisations, laws and scientific hypotheses can gain positive support from empirical evidence. The theory comes in different forms depending on the specific account of how confirmation works.

instrumentalism   The doctrine that scientific theories, and in particular the parts of them that seem to refer to unobservable entities, are merely instruments whose value consists in their ability successfully to predict what can be observed, that is, the outcomes of experiments, rather than in their description of the fundamental structure of reality. Hence, instrumentalism is a form of **antirealism**.

metaphysical realism   The view that our ordinary language refers to, and sometimes says, true things about the world, and that the latter is independent of our minds and cognition.

metaphysics   Metaphysics is that branch of philosophy that studies questions about the fundamental nature of reality, such as what is the nature of space and time, what are laws of nature as opposed to accidental regularities, what fundamental categories of existing things are there, and so on. Scientific realists often think that science, especially physics, can answer metaphysical questions.

methodology   Methodology means the theory of method and in philosophy of science it is the study of techniques and procedures for carrying out experiments, interpreting data and developing and testing theories. Hence, we may talk of the methodology of a particular experiment like a drug trial, or of the methodology of science where we refer to the study of the scientific method in general.

necessary   A proposition is logically necessarily true (is a tautology) if it could not possibly be false (if its negation is a contradiction). Following Leibniz, many modern philosophers equate necessary truth with truth in all possible worlds. We may also speak of a proposition or state of affairs being physically necessary, which just means that it must be so according to the laws of physics. Propositions that are neither necessarily true nor necessarily false are said to be *contingent*.

necessary condition   A necessary condition for the truth of a proposition is a condition that has to be satisfied for the condition to be true; for example, a necessary condition for a geometric shape to be a square is for it to have four sides. Hence, where *P* is a necessary condition for *Q*, we can say *Q* only if *P*, and if *Q* then *P* (something is a square only if it has four sides, and if something is a square then it has four sides). If *P* is a necessary condition for *Q*, then *Q* is a **sufficient condition** for *P*.

negation   The negation of a proposition, *p*, is the proposition expressed by 'it is not the case that *p*' or 'not *p*'. Usually, it is clear what the negation of a proposition is. For example, the negation of 'Socrates died in 399 BC', is 'Socrates did not die in 399 BC'. In other cases, what is the correct way of expressing the negation of a proposition is not so obvious. For example, the negation of 'all swans are white' is not 'all swans are not white', but 'it is not the case that all swans are white', or 'there is at least one non-white swan'.

phenomenalism   The metaphysical thesis according to which there is nothing over and above the phenomena we observe. What we think of as mind-independent objects are in fact logical constructs out of actual and possible sensations.

proposition   A proposition says that a particular state of affairs is the case and may be either true or false. The same proposition may be expressed by more than one sentence, for example, 'snow is white' and 'schnee ist weiss', and the same sentence may express more than one proposition, for example, 'James is at the bank' may refer to the state of affairs where I am by a river, or the state of affairs where I am at the local branch of a financial institution. In classical logic, Aristotle's three laws are assumed according to which everything is identical with itself (the *law of identity*), every proposition is either true or false (the *law of the excluded middle*), and no proposition is both true and false (the *law of non-contradiction*). There are logical systems where one or more of these laws are denied, but it seems fair to say that in the course of everyday reasoning most people operate under the assumption that these laws hold.

realism   In philosophy of science this is the view that we can know that our best scientific theories really refer to unobservable entities that exist independently of our minds. In general philosophical terms, a realist about something is someone who thinks that we can know it exists as a mind-independent entity. However, there are other more precise uses of the term so it is important to pay attention to how the term is being used in a given context.

reductionism   Reductionism is the thesis that the laws and explanations of high level theories in, say, chemistry and biology can be reconstructed entirely in the terms of more fundamental theories in physics.

reductive empiricism   This is a form of antirealism about science according to which theoretical terms can be defined in terms of observational concepts; hence, statements involving them are assertoric. However, according to reductive empiricism scientific theories should not be taken literally as referring to unobservable objects in denial of the semantic component of scientific realism.

semantic instrumentalism   This is a form of antirealism about science according to which the theoretical terms of scientific theories should not be taken literally as referring to unobservable entities, because they are merely logical constructs used as tools for systematising relations between phenomena; theoretical hypotheses are not assertoric.

semantics   In philosophy, this is the study of meaning, reference, truth and other features of language beyond its grammatical and formal structure.

social constructivism   The view that the entities in some domain exist, but are mind-dependent in the sense of not existing over and above our construction of them.

sufficient condition   A sufficient condition for the truth of a proposition is one such that if it is true then the proposition is true; for example, it being Saturday is a sufficient condition for it to be the weekend. Hence, where *A* is a sufficient condition for *B*, we can say, if *A* then *B*, and *A* only if *B* (if it is Saturday then it is the weekend, and it is Saturday only if it is the weekend). If *A* is a sufficient condition for *B*, then *B* is a **necessary condition** for *A*.

teleology   The study of final causes. Aristotle distinguishes between four types of cause, efficient, material, formal and final. For example, consider a statue of Socrates; the efficient cause is the action of the sculptor with a chisel, the material cause is the marble from which the statue is made, the formal cause is the image of Socrates that the sculptor is trying to reproduce, and the final cause is the goal or end of the sculptor in making the statue, perhaps to inspire the members of a philosophy department where the statue will be housed. Hence, final causes are ends or purposes. Since the scientific revolution, many philosophers have held that they ought to have no place in natural science.

Bibliography

—·◎·—

Achinstein, P. (1991) *Particles and Waves*, Oxford: Oxford University Press.

Armstrong, D. (1983) *What is a Law of Nature?*, Cambridge: Cambridge University Press.

Ayer, A. (1940) *The Foundations of Empirical Knowledge*, London: Macmillan.

Ayer, A.J. (1952) *Language Truth and Logic*, Cambridge: Cambridge University Press.

Ayer, A.J. (1956) *The Problem of Knowledge*, Harmondsworth, Middlesex: Penguin.

Barnes, B., Bloor, D. and Henry, J. (1996) *Scientific Knowledge: A Sociological Analysis*, London: Athlone.

Berkeley, G. (1975a) *The Principles of Human Knowledge*, in M.R. Ayers (ed.) *Berkeley Philosophical Works*, London: Everyman.

Berkeley, G. (1975b) *Berkeley Philosophical Works*, London: Everyman.

Boyd, R. (1984) 'The current status of scientific realism', in J. Leplin (ed.) *Scientific Realism*, Berkeley: University of California Press, pp. 41–82.

Boyd, R. (1985) '*Lex Orandi est Lex Credendi*', in P.M. Churchland and C.A. Hooker (eds) *Images of Science*, Chicago: University of Chicago Press, pp. 3–34.

Braithwaite, R. (1953) *Scientific Explanation*, Cambridge: Cambridge University Press.

Carnap, R. (1952) *The Continuum of Inductive Methods*, Chicago: University of Chicago Press.

Carnap, R. (1959) 'The elimination of metaphysics through logical analysis of language', in A.J. Ayer (ed.) *Logical Positivism*, New York: Free Press.

Carroll, L. (1895) 'What the tortoise said to Achilles', *Mind*, 4, pp. 278–80.

Cartwright, N. (1983) *How the Laws of Physics Lie*, Oxford: Oxford University Press.

Cartwright, N., Shomar, T. and Suárez, M. (1995) 'The tool box of science: tools for building of models with a superconductivity example', in

W.E. Herfel, W. Krajewski, W. I. Niiniluoto and R. Wójcicki (eds) *Theories and Models in Science*, Amsterdam: Rodolfi

Churchland, P. (1979) *Scientific Realism and the Plasticity of Mind*, Cambridge: Cambridge University Press.

Churchland, P. (1985) 'The ontological status of observables: in praise of the superempirical virtues', in P. Churchland, and C. Hooker (eds), *Images of Science*, Chicago: University of Chicago Press.

Couvalis, G. (1997) *The Philosophy of Science: Science and Objectivity*, Chapter 1, London: Sage.

Descartes, R. (1941, tr. 1954) *Meditations on First Philosophy*, in E. Anscombe and P. Geach (eds), *Descartes Philosophical Writings*, London: Nelson.

Devitt, M. (1991) *Realism and Truth*, Oxford: Blackwell.

Duhem, P. (1906, tr. 1962) *The Aim and Structure of Physical Theory*, New York: Athenum.

Dupré, J. (1993) *The Disorder of Things: Metaphysical Foundations of the Disunity of Science*, Cambridge, MA: Harvard University Press.

Eddington, A. (1928) *The Nature of the Physical World*, Cambridge: Cambridge University Press.

Feyerabend, P. (1977) *Against Method*, London: New Left Books.

Field, H. (1995) 'Theory change and the indeteminacy of reference', in P. Lipton (ed.) *Theory, Evidence and Explanation*, Aldershot: Dartmouth.

Fine, A. (1984) 'The natural ontological attitude', in J. Leplin (ed.) *Scientific Realism*, Berkeley: University of California Press, pp. 83–107.

Fodor, J. (1984) 'Observation reconsidered', *Philosophy of Science*, **51**, pp. 23–43.

Friedman, M. (1974) 'Explanation and scientific understanding', *Journal of Philosophy*, **LXXI**, pp. 5–19.

Friedman, M. (1999) *Logical Posistivism Reconsidered*, Cambridge: Cambridge University Press.

Glymour, C. (1980) *Theory and Evidence*, Princeton, NJ: Princeton University Press.

Goodman, N. (1973) *Fact, Fiction and Forecast*, Indianapolis: Bobbs-Merrill.

Gower, B. (2000) 'Cassirer, Schlick and "structural" realism: the philosophy of the exact sciences in the background to early logical empiricism', *British Journal for the History of Science*, **8**, pp. 71–106.

Hacking, I. (ed.) (1981) *Scientific Revolution*, Oxford: Oxford University Press.

Hacking, I. (1983) *Representing and Intervening*, Cambridge: Cambridge University Press.

Hanfling, O. (ed.) (1981) *Essential Readings in Logical Positivism*, Oxford: Blackwell.

Hanson, N.R. (1958) *Patterns of Discovery*, Cambridge: Cambridge University Press.

Hardin, C.L. and Rosenberg, A. (1982) 'In defence of convergent realism', *Philosophy of Science*, **49**, pp. 604–15.

Harding, S. (ed.) (1976) *Can Theories be Refuted? Essays on the Duhem–Quine Thesis*, Dordrecht, The Netherlands: D. Reidel.

Harman, G. (1965) 'Inference to the best explanation', *Philosophical Review*, **74**, pp. 88–95.

Hempel, C. (1965) *Aspects of Scientific Explanation*, New York: Free Press.

Hoefer, C. and Rosenberg, A. (1994) 'Empirical equivalence, underdetermination, and systems of the world', *Philosophy of Science*, **61**, pp. 592–607.

Horwich, P. (1982) *Probability and Evidence*, Cambridge: Cambridge University Press.

Horwich, P. (1991) 'On the nature and norms of theoretical commitment', *Philosophy of Science*, **58**, pp. 1–14.

Howson, C. and Urbach, P. (1993) *Scientific Reasoning: The Bayesian Approach*, La Salle, IL: Open Court.

Hoyningen-Huene, P. (1993) *Reconstructing Scientific Revolutions: Thomas Kuhn's Philosophy of Science*, Chicago: University of Chicago Press.

Hume, D. (1963) *An Enquiry Concerning Human Understanding*, La Salle, IL: Open Court.

Hume, D. (1978) *A Treatise of Human Nature*, Oxford: Oxford University Press.

Jones, R. (1991) 'Realism about what?', *Philosophy of Science*, **58**, pp. 185–202.

Kitcher, P. (1993) *The Advancement of Science: Science without Legend, Objectivity without Illusions*, Oxford: Oxford University Press.

Kuhn, T.S. (1957) *The Copernican Revolution: Planetary Astronomy in the Development of Western Thought*, Cambridge, MA: Harvard University Press.

Kuhn, T.S. (1962, 2nd edn 1970) *The Structure of Scientific Revolutions*, Chicago: University of Chicago Press.

Kuhn, T.S. (1977) *The Essential Tension*, Chicago: University of Chicago Press.

Kukla, A. (1993) 'Laudan, Leplin, empirical equivalence and underdetermination', *Analysis*, **53**, pp. 1–7.

Kukla, A. (1996) 'Does every theory have empirically equivalent rivals?', *Erkenntnis*, **44**, pp. 137–66.

Kukla, A. (1998) *Studies in Scientific Realism*, Oxford: Oxford University Press.

Kukla, A. (2000) *Social Constructivism and the Philosophy of Science*, London: Routledge,

Ladyman, J. (1998) 'What is structural realism?', *Studies in History and Philosophy of Science*.

Ladyman, J. (2000) 'What's really wrong with constructive empiricism?:

van Fraassen and the metaphysics of modality', *British Journal for the Philisophy of Science*, **51**, pp. 837–56.

Ladyman, J., Douven, I., Horsten, L. and van Fraassen, B.C. (1997) 'In defence of van Fraassen's critique of abductive reasoning: a reply to Psillos', *Philosophical Quarterly*, **47**, pp. 305–21.

Lakatos, I. (1968) 'Criticism and the methodology of scientific research programmes', *Proceedings of the Aristotelian Society*, **69**, pp. 149–86.

Lakatos, I. and Musgrave, A. (eds) (1970) *Criticism and the Growth of Knowledge*, Cambridge: Cambridge University Press.

Laudan, L. (1977) *Progress and its Problems*, Berkeley: University of California Press.

Laudan, L. (1981) 'A confrontation of convergent realism', *Philosophy of Science*, **48**, pp. 19–48.

Laudan, L. (1984) *Science and Values*, Berkeley: University of California Press.

Laudan, L. and Leplin, J. (1991) 'Empirical equivalence and underdetermination', *Journal of Philosophy*, **88**, pp. 269–85.

Laudan, L. and Leplin, J. (1993) 'Determination underdeterred', *Analysis*, **53**, pp. 8–15.

Leplin, J. (1997) *A Novel Defense of Scientific Realism*, Oxford: Oxford University Press.

Lipton, P. (1991) *Inference to the Best Explanation*, London: Routledge.

Locke, J. (1964) *An Essay Concerning Human Understanding*, Glasgow: Collins.

McCulloch, G. (1989) *The Game of the Name: Introducing Logic, Language and the Mind*, Chapters 2 and 3, Oxford: Oxford University Press.

Maxwell, G. (1962) 'The ontological status of theoretical entities', in H. Feigl and G. Maxwell (eds) *Minnesota Studies in the Philosophy of Science*, Volume 3, Minneapolis: University of Minnesota Press, pp. 3–14.

Merton, R.K. (1973) *The Sociology of Science*, Chicago: University of Chicago Press.

Musgrave, A. (1992), 'Discussion: realism about what?', *Philosophy of Science*, **59**, pp. 691–7.

Musgrave, A. (1993), *Common Sense, Science and Scepticism: A Historical Introduction to the Theory of Knowledge*, Cambridge: Cambridge University Press.

Nagel, E. (1961) *The Structure of Science: Problems in the Logic Scientific Explanation*, New York: Harcourt, Brace and World.

Newton-Smith, W. (1981) *The Rationality of Science*, London: Routledge.

Papineau, D. (1979) *Theory and Meaning*, Oxford: Oxford University Press.

Papineau, D. (1993) *Philosophical Naturalism*, Oxford: Blackwell.

Papineau, D. (ed.) (1996) *Philosophy of Science*, Oxford: Oxford University Press.

Putnam, H. (1975a) *Mathematics, Matter and Method: Philosophical Papers*, Volume 1, Cambridge: Cambridge University Press.

Putnam, H. (1975b) *Mind, Language and Reality: Philosophical Papers*, Volume 2, Cambridge: Cambridge University Press.

Planck, M. (1996) 'The universe in the light of modern physics', in W. Schirmacher (ed.) *German Essays on Science in the 20th Century*, New York: Continuum, pp. 38–57.

Poincaré, H. ([1905] 1952) *Science and Hypothesis*, New York: Dover.

Popper, K. ([1934] 1959) *The Logic of Scientific Discovery*, London: Hutchinson.

Popper, K. (1969) *Conjectures and Refutations*, London: Routledge and Kegan Paul.

Psillos, S. (1994) 'A philosophical study of the transition from the caloric theory of heat to thermodynamics: resisting the pessimistic meta-induction', *Studies in the History and Philosophy of Science*, **25**, pp. 159–90.

Psillos, S. (1996) 'On van Fraassen's critique of abductive reasoning', *Philosophical Quarterly*, **46**, pp. 31–47.

Psillos, S. (1999) *Scientific Realism: How Science Tracks Truth*, London: Routledge.

Quine, W.v.O. (1953) 'Two dogmas of empiricism', in *From a Logical Point of View*, Cambridge, MA: Harvard University Press.

Redhead, M. (1996) 'Quantum field theory and the philosopher', offprint.

Rosen, G. (1994) 'What is constructive empiricism?', *Philosophical Studies*, **74**, pp. 143–78.

Ruben, D. (1990) *Explaining Explanation*, London: Routledge.

Ruben, D. (ed.) (1993) *Explanation*, Oxford: Oxford University Press.

Russell, B. (1912) *The Problems of Philosophy*, Chapter 6, Oxford: Oxford University Press.

Salmon, W. (1984) *Scientific Explanation and the Casual Structure of the World*, Princeton, NJ: Princeton University Press.

Shapere, D. (1981) 'Meaning and scientific change', in I. Hacking (ed.) *Scientific Revolutions*, Oxford: Oxford University Press.

Shapiro, S. (2000) *Thinking about Mathematics*, Chapter 5, Oxford: Oxford University Press.

Sklar, L. (1974) *Space, Time and Spacetime*, Berkeley: University of California Press.

Smart, J. (1963) *Philosophy and Scientific Realism*, London: Routledge.

Stein, H. (1989) 'Yes, but ... some skeptical remarks on realism and antirealism', *Dialectica*, **43**, pp. 47–65.

Swinburne, R. (ed.) (1974) *Justification of Induction*, Oxford: Oxford University Press.

Van Fraassen, B.C. (1980) *The Scientific Image*, Oxford: Oxford University Press.

Van Fraassen, B.C. (1985) 'Empiricism in the philosophy of science', in

P. Churchland and C. Hooker (eds) *Images of Science*, Chicago: University of Chicago Press.

Van Fraassen, B.C. (1989) *Laws and Symmetry*, Oxford: Oxford University Press.

Van Fraassen, B.C. (1994) 'Against transcendental empricism', in T.J. Stapleton (ed.) *The Question of Hermeneutics*, Amsterdam: Kluwer, pp. 309–35.

Woolhouse, R.S. (1988) *The Empiricists*, Chapter 8, Oxford: Oxford University Press.

Worrall, J. (1984) 'An unreal image', review article of van Fraassen (1980), *British Journal for the Philosophy of Science*, 35, pp. 65–79.

Worrall, J. (1989) 'Structural realism: the best of both worlds?', *Dialectica*, 43, pp. 99–124.

Worrall, J. (1994) 'How to remain (reasonably) optimistic: scientific realism and the "luminiferous ether"', in D. Hull, M. Forbes and R.M. Burian (eds) *P.S.A. 1994*, Vol. 1., Philosophy of Science Association, pp. 334–42.

Index

—◦◯◦—

276